# Environmental Life Cycle Costing

## Other Titles from the Society of Environmental Toxicology and Chemistry (SETAC)

*Valuation of Ecological Resources: Integration of Ecology and Socioeconomics in Environmental Decision Making*
Stahl, Kapustka, Munns, Bruins, editors
2007

*Genomics in Regulatory Ecotoxicology: Applications and Challenges*
Ankley, Miracle, Perkins, Daston, editors
2007

*Population-Level Ecological Risk Assessment*
Barnthouse, Munns, Sorensen, editors
2007

*Effects of Water Chemistry on Bioavailability and Toxicity of Waterborne Cadmium, Copper, Nickel, Lead, and Zinc on Freshwater Organisms*
Meyer, Clearwater, Doser, Rogaczewski, Hansen
2007

*Ecosystem Responses to Mercury Contamination: Indicators of Change*
Harris, Krabbenhoft, Mason, Murray, Reash, Saltman, editors
2007

*Genomic Approaches for Cross-Species Extrapolation in Toxicology*
Benson and Di Giulio, editors
2007

*New Improvements in the Aquatic Ecological Risk Assessment of Fungicidal Pesticides and Biocides*
Van den Brink, Maltby, Wendt-Rasch, Heimbach, Peeters, editors
2007

*Freshwater Bivalve Ecotoxicology*
Farris and Van Hassel, editors
2006

*Estrogens and Xenoestrogens in the Aquatic Environment: An Integrated Approach for Field Monitoring and Effect Assessment*
Vethaak, Schrap, de Voogt, editors
2006

For information about SETAC publications, including SETAC's international journals, *Environmental Toxicology and Chemistry* and *Integrated Environmental Assessment and Management*, contact the SETAC Administratice Office nearest you:

SETAC Office
1010 North 12th Avenue
Pensacola, FL 32501-3367 USA
T 850 469 1500 F 850 469 9778
E setac@setac.org

SETAC Office
Avenue de la Toison d'Or 67
B-1060 Brussells, Belguim
T 32 2 772 72 81 F32 2 770 53 86
E setac@setaceu.org

***www.setac.org***

**Environmental Quality Through Science®**

# Environmental Life Cycle Costing

*Edited by*
**David Hunkeler, Kerstin Lichtenvort, and Gerald Rebitzer**

*Lead Authors*

**Andreas Ciroth**
**Gjalt Huppes**
**Walter Klöpffer**
**Ina Rüdenauer**
**Bengt Steen**
**Thomas Swarr**

*Coordinating Editor of SETAC Books*
Joseph W. Gorsuch
Gorsuch Environmental Management Services, Inc.
Webster, New York, USA

CRC Press is an imprint of the
Taylor & Francis Group, an **informa** business

Published in collaboration with the Society of Environmental Toxicology and Chemistry (SETAC)
1010 North 12th Avenue, Pensacola, Florida 32501
Telephone: (850) 469-1500 ; Fax: (850) 469-9778;
Email: setac@setac.org
Web site: www.setac.org
ISBN: 1-880611-38-X (SETAC Press)


CRC Press is an imprint of Taylor & Francis Group, an Informa business
SETAC Press is an imprint of the Society of Environmental Toxicology and Chemistry (SETAC)


Printed in the United States of America on acid-free paper
10 9 8 7 6 5 4 3 2 1

International Standard Book Number-13: 978-1-880611-83-8 (Hardcover; CRC Press)
International Standard Book Number-13: 978-1-58488-661-7 (Hardcover; SETAC)



**Library of Congress Cataloging-in-Publication Data**

Environmental life cycle costing / editors, David Hunkeler, Kerstin Lichtenvort, and Gerald Rebitzer.
p. cm.
Includes bibliographical references and index.
ISBN 978-1-4200-5470-5 (alk. paper)
1. Environmental engineering. 2. Environmental engineering--Economic aspects. 3. Life cycle costing. I. Hunkeler, David (David Jerome), 1962- II. Lichtenvort, Kerstin. III. Rebitzer, Gerald. IV. SETAC-Europe. V. Title.

TA170.E635 2008
628--dc22 2008003600

**Visit the Taylor & Francis Web site at**
**http://www.taylorandfrancis.com**

**and the CRC Press Web site at**
**http://www.crcpress.com**

**and the SETAC Web site at**
**www.setac.org**

# SETAC Publications

Books published by the Society of Environmental Toxicology and Chemistry (SETAC) provide in-depth reviews and critical appraisals on scientific subjects relevant to understanding the impacts of chemicals and technology on the environment. The books explore topics reviewed and recommended by the Publications Advisory Council and approved by the SETAC North America, Latin America, or Asia/Pacific Board of Directors; the SETAC Europe Council; or the SETAC World Council for their importance, timeliness, and contribution to multidisciplinary approaches to solving environmental problems. The diversity and breadth of subjects covered in the series reflect the wide range of disciplines encompassed by environmental toxicology, environmental chemistry, and hazard and risk assessment, and life-cycle assessment. SETAC books attempt to present the reader with authoritative coverage of the literature, as well as paradigms, methodologies, and controversies; research needs; and new developments specific to the featured topics. The books are generally peer reviewed for SETAC by acknowledged experts.

SETAC publications, which include Technical Issue Papers (TIPs), workshops summaries, newsletter (SETAC Globe), and journals (Environmental Toxicology and Chemistry and Integrated Environmental Assessment and Management), are useful to environmental scientists in research, research management, chemical manufacturing and regulation, risk assessment, and education, as well as to students considering or preparing for careers in these areas. The publications provide information for keeping abreast of recent developments in familiar subject areas and for rapid introduction to principles and approaches in new subject areas.

SETAC recognizes and thanks the past coordinating editors of SETAC books:

A.S. Green, International Zinc Association
Durham, North Carolina, USA

C.G. Ingersoll, Columbia Environmental Research Center
US Geological Survey, Columbia, Missouri, USA

T.W. La Point, Institute of Applied Sciences
University of North Texas, Denton, Texas, USA

B.T. Walton, US Environmental Protection Agency
Research Triangle Park, North Carolina, USA

C.H. Ward, Department of Environmental Sciences and Engineering
Rice University, Houston, Texas, USA

# Contents

# List of Figures

# List of Tables

# About the Editors

**David Hunkeler** is the general director of AQUA+TECH, in Geneva Switzerland, a firm in its tenth year of operations, focusing on water treatment. AQUA+TECH is the smallest global flocculant producer and assists its clients with optimizing the life cycle costs and environmental impacts of their clarification choices. They work principally in Africa and Asia. AQUA+TECH was voted the Europe's top environmental firm, in 2002, by the Wall St. Journal Europe. It also received the Bronze Medal for business innovation by the same organization as received the Swiss Economic Award in 2003. Hunkeler has received the de Vigier prize as Switzerland's top young entrepreneur. He has authored over 200 peer reviewed publications, 100 poems, and 2 books, and he has been a coeditor on five monographs. He is the cofounder of the Foundation InsuLéman, a non-profit organization for the treatment of immune deficient diseases via cell transplantation and is a proud member of the ChicagoProject, which works towards a clinical therapy for diabetes. Five of his former collaborators are now professors, while six are entrepreneurs.

**Kerstin Lichtenvort** holds a PhD in environmental engineering from the Technical University Berlin and has been active in life cycle assessment research since 1995. She started the work on this book in the SETAC working group on life cycle costing, while being coordinator of the grEEEn project, which established a cost management system for greening electrical and electronic equipment. The project, funded by the Fifth Framework Program for Research and Technological Development of the European Commission, discussed the basic principles of Environmental Life Cycle Costing with key stakeholders from industry, research, and consultancy. In the Department for Systems Environmental Engineering at the Institute for Environmental Technology of the Technical University Berlin, Lichtenvort developed strategies and tools in the field of energy, environmental, economic, and risk assessments; including ecodesign schemes, energy and transport trends, and strategies to tackle climate change and resource depletion. Nowadays, she works in Brussels for the Intelligent Energy Europe Program on the field of energy efficiency, energy using products, and ecodesign.

**Gerald Rebitzer**, currently director product stewardship of Alcan Packaging, holds a PhD in life cycle management from the Swiss Federal Institute of Technology, Lausanne and a master's in environmental engineering from the Technical University Berlin. Besides his Alcan Packaging internal tasks, which focus on the implementation of product sustainability in the multinational's business processes and leveraging sustainability in cooperation with suppliers, customers, and other stakeholders, Rebitzer is editor for *The International Journal of Life Cycle Assessment*, lecturer at the ETH Zurich, reviewer for the European Commission, and leads related work in SETAC and UNEP/SETAC as well as the Word Business Council for Sustainable Development, among other activities. Recently, he co-organized LCM2007, the largest product related sustainability conference to date. Rebitzer has published more than 100 papers and book chapters in his area of expertise.

# Contributors

**Andreas Ciroth**
GreenDeltaTC GmbH
Berlin, Germany

**Carl-Otto Gensch**
Öko-Institut e.V.
Freiburg, Germany

**Edeltraud Günther**
Faculty of Business and Economics
Technische Universitaet Dresden
Dresden, Germany

**Andrea Heilmann**
Hochschule Harz
Wernigerode, Germany

**Holger Hoppe**
Faculty of Business and Economics
Technische Universitaet Dresden
Dresden, Germany

**David Hunkeler**
AQUA+TECH
Geneva, Switzerland

**Gjalt Huppes**
Institute for Environmental Sciences (CML)
Leiden University
Leiden, The Netherlands

**Walter Klöpffer**
LCA Consult & Review
Frankfurt, Germany

**Kerstin Lichtenvort**
Brussels, Belgium

**Kjerstin Ludvig**
Akzo Nobel
Amsterdam, The Netherlands

**Shinichiro Nakamura**
Graduate School of Economics
Waseda University
Tokyo, Japan

**Bruno Notarnicola**
Department of Commodity Science
University of Bari
Bari, Italy

**Andrea Pelzeter**
Fachhochschule für Wirtschaft Berlin
Fachbereich Berufsakademie
Fachrichtung Facility Management
Berlin, Germany

**Martina Prox**
ifu Hamburg
Hamburg, Germany

**Gerald Rebitzer**
Alcan Packaging
Neuhausen, Switzerland

**Ina Rüdenauer**
Öko-Institut e.V.
Freiburg, Germany

**Wulf-Peter Schmidt**
Ford Werke GmbH
Cologne, Germany

**Stefan Seuring**
University of Kassel
Kassel and Witzenhausen, Germany

**Ernst Spindler**
Vinnolit GmbH & Co. KG
Burghausen, Germany

**Bengt Steen**
Department of Environmental Systems Analyses
Chalmers University of Technology
Göteborg, Sweden

**Thomas Swarr**
Adjunct, Rensselaer
Hartford, CT, USA

**Christian Trescher**
Die Ingenieurwerkstatt Gesellschaft für Lifecycle-Engineering mbH
Wiesbaden, Germany

**Karli Verghese**
RMIT University
Melbourne, Australia

# Preface: About This Book

This book summarizes 3 years of deliberations of the SETAC-Europe Working Group on Life Cycle Costing, followed by 2 years of writing. The working group, with approximately 20 actively participating members and another 5 dozen corresponding scientists, consultants, and businesspersons, had its kickoff meeting in December 2002 in Barcelona at the Society of Environmental Toxicology and Chemistry (SETAC) Life Cycle Assessment (LCA) Case Studies Symposium, with its last retreat in Barcelona in September 2005 during the Life Cycle Management International Conference (LCM2005). In between, the members met 5 times at SETAC meetings in Hamburg, Prague, Lausanne, Bologna, and Lille.

The initial charge of the working group was mandated by the SETAC-Europe LCA steering committee, which envisioned the need to develop and formalize sustainability assessments. Specifically, standardized methods for economic as well as societal assessments were seen as another 2 essential pillars of a sustainability assessment, in addition to the now well-established LCA methodology satisfying the requirements for environmental analyses. Environmental LCC, as detailed in this book and as a precursor to a code of practice, is a method complementary to LCA, utilizing equivalent system boundaries and functional units, and can be seen as a 2nd pillar of the sustainability assessment of products (including services). The evolving societal assessment, as noted by Walter Klöpffer, editor of *The International Journal of Life Cycle Assessment* (Klöpffer 2003), then represents the 3rd pillar.

This book not only focuses on the aforementioned environmental LCC but also analyzes a related, established, cost management-based method stemming primarily from public and military origins, which is termed "conventional LCC." Additionally, a more expanded form of costing was identified in surveys conducted at the outset of the working group's activities, and it includes costs external to the economic system, at least at the present. This evolving method is termed "societal LCC," and it also is presented herein, to the extent of its present development. The latter 2 approaches are included in order to put environmental LCC into context and also to show the relationship to other concepts.

This book defines these 3 variants of life cycle costing, providing case studies for each. Throughout each chapter, case study boxes demonstrate the process for carrying out an LCC, from problem definition to analysis and ultimate presentation to the decision maker. One case, for an idealized washing machine, is used throughout. By using this case, the authors and editors hope that readers will understand not only the nuances of LCC but also the means to carry it out and to benchmark new LCC concepts and studies against existing procedures and examples.

We hope that the following dialog will help readers to develop their initial and/or fundamental understanding of life cycle costing.

**David Hunkeler, Kerstin Lichtenvort, and Gerald Rebitzer**
July 2007

# A Dialog, over Coffee, about Life Cycle Costing

Josef, a 50-something European businessman, is met at New York's LaGuardia Airport by his American colleague, Fathima. Following a short bus ride, they find themselves across the street from Grand Central Station with time for coffee and pastry at an Italian bistro.

*Fathima (looking at Josef for confirmation, receiving a nod, then pointing at the display case with a revealing smile):* 2 coffees and 2 cannoli.

*Josef:* Looks lovely, albeit maybe a bit decadent for 3 in the afternoon, though the long flight has me looking forward to both.

*Ines (an afternoon regular on the next stool, watching Josef drink his espresso in one sip and hesitating to begin cutting into the wafer roll):* Bit funny that the coffee is served in cups and dessert on a disposable paper plate.

*Gigi (the owner, a gregarious and tanned Mediterranean, introducing himself and gesticulating with his hands):* It is cheaper that way. We turn over so much coffee, and it's drunk so rapidly here at the bar, that the best way is to wash about 40 of the demitasse cups at a time. The customers like the ceremony, too. Plates, though, require too large a machine to wash, and their disposal does not generate too many garbage bags that we have to pay to cart away, so we use paper.

*Josef (first to Fathima, though not out of range of Gigi):* How could it be that it is less expensive to wash one — the cup — but not the other — the plate?

*Fathima (midsentence, looking for a confirmation from Gigi):* If you calculate the cost of the energy and detergent required to wash a cup, for example — perhaps he adds in labor cost as well — then you need to use the cup about 1000 times before its use is as inexpensive as that of a disposable Styrofoam alternative. Gigi probably gets at least 2000 uses before the china cup breaks, so he uses china and saves money.

*Josef (understanding but with the look that he is still either doing the math or trying to figure out the point, and looking at Gigi):* Could be, and I have no reason to not have complete confidence in you. However, why, then, do you dispose of the plates?

Gigi and Fathima both begin to speak, though ultimately the 2nd-generation immigrant from Benvenuto bows slightly and leaves way to his female regular.

*Ines:* The paper plates are thin and elegant. People like them, and Gigi picks the color to suit your snack. The plates come with different patterns for different times of the year, and they're dainty. This one is red and green for Christmas! Gigi can dispose of a day's worth in 3 small bags for a cost of $6.

*Gigi (continuing):* Yes, and one needs a — let's say, a big — dishwasher to hold all the plates. My electricity bill is already $300 per month. With an extra big appliance, it would cost me $100 more. Even without counting salaries, I save $20 a week using the paper plates.

Josef finishes his cannelloni, buys some torrone from the counter, and pays the bill, leaving the change and bringing out his portable.

*Josef:* Amazing things these are, though I am more addicted to them than either caffeine or sweets (he admits this without too much remorse, though, obviously, with some reflection). So, the china cups cost $1000 per month less than Styrofoam cups if you wash them, and you figure that the energy, the soap, and Rafaela's time are not that costly. (Gigi smiles at all of them as his daughter clears off the countertop, nodding a confirmation.)

*Fathima:* It's all about relative product comparisons. The difference between 2 alternatives can be calculated a lot easier than an absolute calculation. Works in cases like this and in full-scale studies, like the one we will work on this afternoon.

*Josef (continuing her sentence almost without a pause):* And, if I understand correctly, the paper plates are less expensive than the china ones over the course of their use and disposal cycle.

*Fathima (in full affirmation):* Yes, because Gigi is not penalized for disposing of paper plates and because the energy needed to wash big china plates is high, not to mention the need for more storage space.

*Gigi (laughing):* Fathima tells me, I think with a life cycle perspective!

*Josef (with an air of obvious, but slight, disagreement):* Whatever! The life cycle of a product lasts from its development through its production, and as it becomes mature in the market, its sales decline. Like a black-and-white television. Everyone who ever passed through a business school — or anyone who even reads the financial papers — would tell you that.

*Fathima (kissing her friend Rafaela once on each cheek as they walk out, and handing Josef his portfolio):* Yes, but life cycle costing is something else. It relates the real money flows to a product from the time its material components are extracted to its manufacture, transport, use, and ultimately its disposal.

*Egido (the slightly round 8-year-old grandson of Gigi, running to give Fathima her purse with some sugar-covered almonds stuck in it):* Zia (Aunt) Fathima, don't forget to tell your friend he can read about it in your book.

*Fathima (hands Michel a pen out of Josef's agenda and rubs his head, replacing the pen with a copy of the book, Environmental Life Cycle Costing):* It's like the coffee and dessert. Gigi buys soap, china cups, and paper plates, and he pays his rent, electricity, and water and his family's salary. He pays to dispose of some of the garbage bags, and he has some transport bills. Those are all costs — real monetary transactions — not just of a product while it is here in the shop but also during its entire life cycle.

*Josef (unable to leave the last word to his friend, leaning his head sideways and waving off a taxi, and summarizing):* Yes, all the costs a product "sees" from the time it is being built to the time it is being buried, so to speak.

*Fathima (walking through 1 of the 2 glass doors of the building, just in time to announce herself, and smiling):* If you tell them in your meeting it's a "cradle-to-grave" analysis, they will know you have become a real expert — and you won't have to read the book in the elevator, though it will be great entertainment on the flight home.

# Executive Summary

**Environmental Life Cycle Costing**

Environmental life cycle costing (LCC) summarizes all costs associated with the life cycle of a product that are directly covered by 1 or more of the actors in that life cycle (e.g., supplier, producer, user or consumer, and those involved at the end of life [EoL]); these costs must relate to real money flows. Externalities that are expected to be internalized in the decision-relevant future comprise real money flows as well, and they must also be included. A complementary life cycle assessment (LCA), with equivalent system boundaries and functional units, is also required. Environmental LCC is performed on a basis analogous to that of LCA, with both being steady-state in nature.

This executive summary defines the principal types of LCC, puts environmental LCC into perspective, and presents an example of their calculation for an idealized case study on washing machines.

## 0.1 THREE CATEGORIES OF LIFE CYCLE COSTING

Conventional life cycle costing (LCC) is, to a large extent, the historic and current practice in many governments and firms, and is based on a purely economic evaluation, considering various stages in the life cycle. It is a quasi-dynamic method and generally includes (conventional) costs associated with a product* that are borne directly by a given actor and is usually presented from the perspective of the producer or consumer alone. Internalized (or to-be-internalized) external costs that are not immediately tangible, or not borne directly by 1 of the life cycle actors in question, are often neglected. Additionally, conventional LCC does not always consider the complete life cycle; for example, EoL operations are not included in any case. In this sense, conventional LCC can be less comprehensive in scope than systematic environmental analyses such as LCA. Conventional LCC usually involves discounted costs. The lowest discounted rate to be applied would be the market interest rate corrected for inflation, or the cost of equity, with the upper range being the internal rate used by organizations for their intended return on investment. The choice of discount is left to the decision maker, as in managerial cost management.

Environmental LCC uses system boundaries and functional units equivalent to those of LCA and is based on the same product system model, addressing the complete life cycle. In this sense, the 2 analyses are seen as complementary in that all costs are included as directly borne throughout the chain, including the already internalized cost of external effects, and those expected to be internalized within a

* Throughout this book, the term "product" is used in the broader context (i.e., also including services; see also ISO 14040/44 2006).

perspective relevant to the decision at hand. These costs are complemented by an LCA for the same product system. Costs that were previously externalities, though, are now internalized in monetary units, and there is no conversion from environmental measures to monetary measures, or vice versa. There should be no double counting of externalities or environmental impacts in LCC and the complementary LCA.

Single-score indicators are not employed, and the ultimate results are "portfolio presentations" of the life cycle costs combined with key environmental life cycle impacts. For example, a portfolio presentation could involve the life cycle costs, in euro per unit, as related to the global warming potential (GWP) in kilograms of $CO_2$ equivalents per unit, should GWP be a validated, above-threshold, environmental impact for the case at hand. Environmental LCC usually is a steady-state method, as is the complementary LCA. Discounting of the final result of environmental LCC specifications is, in this case, not consistent, nor is it easily carried out. However, the usual discounting of cash flows is the norm.

Societal LCC, as developed for cost–benefit analysis (CBA), uses an expanded macroeconomic system and includes a larger set of costs, including those that will be, or could be, relevant in the long term for all stakeholders directly affected and for all indirectly affected through externalities. A key difference between societal LCC and both conventional and environmental LCC is the encompassing nature of the stakeholder group, which includes governments and other public bodies not directly concerned with the product system. Therefore, the costs involved do not include transfer payments, such as subsidies and taxes, because they are internal to the system. Societal LCC includes the (not necessarily) monetized environmental effects of the investigated product as may be based on a complementary LCA. This quite comprehensive LCC applies a social rate of time preference type of discounting. Aligning to the Brundtland Commission's requirements (1987), this approach would lead to an approximate value of below 0.1% per year, though other societal preferences may be used. As is the case for conventional LCC, the method is quasi-dynamic. If all externalities are monetized and all transfer payments are subtracted, underestimation or double counting of societal costs is avoided. Societal LCC aims to include all environmental (and social) effects and thus is a stand-alone method not accompanied by an LCA or an additional social assessment.

This book is titled *Environmental Life Cycle Costing* as this type of LCC presents a new LCC methodology that is consistent with LCA and suitable as a pillar of sustainability, going beyond the well-known conventional LCC and factoring out still disputable assumptions on societal LCC.

## 0.2 SYSTEM BOUNDARIES IN ENVIRONMENTAL LIFE CYCLE COSTING

The conceptual framework of environmental LCC is based on the physical product life cycle of LCA, as explained in Figure 0.1. Also, the possible relationship of LCC and LCA to societal assessments (focusing on social aspects), for example, including employment conditions and employment volumes, may be placed in this framework.

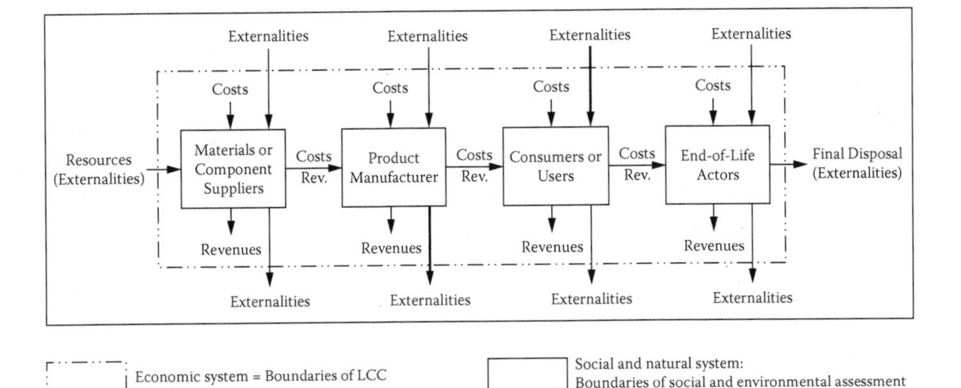


**FIGURE 0.1** Conceptual framework of environmental LCC. *Source*: Rebitzer and Hunkeler (2003).

One can differentiate, in Figure 0.1, between the following:

- Internal costs along the life cycle of a product, with internal implying that an actor (i.e., a producer, transporter, consumer, or other directly involved stakeholder) is paying for the production, use, or EoL expenses, and thereby the internal costs can be connected to a business cost. This concerns all the costs and revenues within the economic system (inside the dashed lines, as represented in Figure 0.1), with costs and revenues of flows between processes netting to 0.
- External costs include the environmental and social impacts not directly borne by any of those taking part in the product life cycle, such as the firms, consumers, or government bodies that are producing, using, or handling the product. These are the so-called externalities, which are outside the economic considerations of those involved in the system, occurring in or through the natural and social systems. Externalities may be monetized hypothetically, as, for example, through willingness-to-pay methods, though in combination with LCA externalities they usually are not monetized because they are covered by the LCA framework. Some (former) externalities are internalized by governments pricing them, as 1 option for the polluter-pays principle. These externalities then lead to real money flows, and to real costs for those causing them, as in several countries is the case with $CO_2$ and $SO_x$ emissions. External cost, therefore, may refer to cost induced to others outside the system boundaries, but internalized as real money flows, and to the cost of externalities as monetized hypothetically.

In this context, it is important to note that the terms and boundaries for economic as well as social and natural systems are not necessarily synonymous with those of the "product system" in LCA (ISO 14040/44 2006), as is detailed in Chapter 3. Therefore, the discussion of externalities and internalization is a key point for

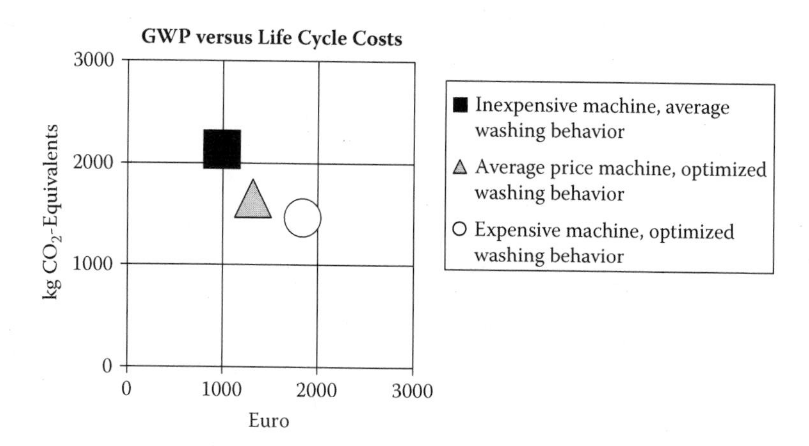


**FIGURE 0.2** Environmental LCC portfolio presentation of 3 alternative washing machines.

environmental LCC, though it differs from societal LCC (Rebitzer et al. 2004), the latter possibly leading to a fully aggregated assessment of environmental and economic (and social) aspects. In environmental LCC, however, nonmonetary units are not converted to monetary values because they are accounted for in environmental terms, as via impact assessment indicator results according to ISO 14040/44 (2006) or via separate social (life cycle) assessments. Societal LCC usually includes the monetization of externalities. However, societal LCC does not (nor should it) substitute for sustainability assessment. Indeed, LCC is seen as 1 of the 3 independent pillars of sustainability analysis, as will be discussed in Chapters 5 and 9. Sustainability assessments, which are still in their infancy, should include LCA, LCC, and a societal assessment, referring to environmental, economic, and social aspects respectively. While the LCA is well developed and consensus is building around LCC methods, comprehensive societal assessments are rather new, and therefore one could reasonably expect their development and standardization to require the next decade to fulfill.

There must be a financial transaction within the value chain for inclusion in environmental and conventional LCC (the concept of real money flows). This inclusion implies that a combination of LCC and LCA will be required to present the results (see portfolio presentations in Figure 0.2).

## 0.3 EXAMPLE CALCULATIONS IN ENVIRONMENTAL LIFE CYCLE COSTING

This book uses an idealized washing machine example as a cross-cutting case throughout all chapters, presented in 10 case study boxes. The example is based on European washing machine technology and wash habits, with the exception of Case Study 8, which is based on a study on Japanese washing machines. The reader should keep in mind that there are differences in washing machine technology and consumer habits (summarized in the Appendix).

The functional unit of the European washing machine is 1840 washing cycles for an average household of 3 people over 11 years. While the details of the methodology and the details of the specific LCC calculations will be left for other chapters, herein the differences in the 3 types of LCC for this case are summarized. This summary should permit the reader and the user to identify the most appropriate form of LCC for their case in question, and to understand the quite significant differences in the calculations. Table 0.1 lists the results of the conventional, environmental, and societal LCC for the washing machine case study. One can observe the following tendencies:

- The life cycle costs increase as one moves from conventional LCC (1172 € per unit), to environmental LCC (1216 € per unit), to societal LCC (1791 € per unit), reflecting the different scopes of assessment and the inclusion of additional monetized "externalities," using fictitious values for monetizing the different environmental effects. This is particularly noticeable in the use stage, though to a lesser extent also in the research and development, preproduction, and production stages and in the disposal stage. These differences also might be much larger or smaller for other applications, so generalization of these differences has to be avoided.
- The impact assessment results, presented in terms of the 5 main impact categories that constitute more than 80% of the total burden, accompany environmental LCC by definition, and are absent from conventional and societal LCC. The dominant impact in this example is clearly the GWP, with the use phase, therefore, of particular concern for improvement.

The presentation of the results for an ultimate decision maker would be, for conventional and societal LCC, in the form of a table such as that presented in this executive summary (see Table 0.1). For the environmental LCC, the results could be a portfolio presentation (Figure 0.2), plotting the overall LCC against, in each figure, 1 LCA result. Figure 0.3 demonstrates that, while the hotspots are similar for the economic (LCC) and environmental (LCA) pillars of sustainability, the latter exemplified by

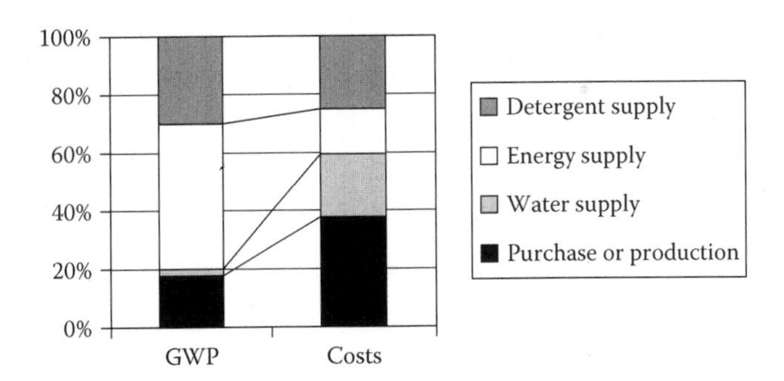


**FIGURE 0.3** Hotspot identification (LCA and environmental LCC) for an average conventional washing machine. *Note*: Costs are in euros, and "GWP" refers to global warming potential in kg of $CO_2$ equivalents.

**TABLE 0.1**

**Results of conventional, environmental, and societal life cycle costing for an idealized washing machine case**

| | Life cycle costs (€ per unit)[a] | | | | | Life cycle impact assessment (LCIA; per unit) | | | | |
|---|---|---|---|---|---|---|---|---|---|---|
| Type of life cycle costing | Research and development | Preproduction | Production | Use and maintenance | End of life (with revenues) | Global warming $CO_2$ equivalent (kg) | Acidification $SO_2$ equivalent (kg) | Eutrophication N equivalent (kg) | Human toxicity benzene equivalent (kg) | Resource depletion oil equivalent (kg) |
| Conventional LCC[b] | 314 | 858 | — | — | — | — | — | — | — | — |
| Environmental LCC | 20 | 216 | 106 | 916 | –42 | 1657 | 8 | 2 | 0.001 | 830 |
| Societal LCC[c] | 445 | 1380 | –34 | — | — | — | — | — | — | — |

[a] The cost per unit could contain other costs for stakeholders other than the ones mentioned (e.g., for public bodies taking care of infrastructure).

[b] Conventional LCC does not, generally, include a complementary LCA because environmental impacts are not considered obligatory in conventional LCC. However, in contrast to this example, conventional LCC may include all life cycle phases if borne by the main actors, including the end of life (e.g., when there are obligatory take-back schemes).

[c] Societal LCC does not, generally, include a complementary LCA, to avoid double counting of environmental impacts (monetary and LCIA).

the GWP score, important differences exist. For example, while the detergent use contributes approximately one-quarter to both the cost and the GWP, energy has a much larger impact share environmentally than economically. The contrary is the case for water use and production phases, where the cost has a larger percentage than the global warming score. The details of the presentation of the LCC results, and the differences between the methods, as well as reasons for their development are elaborated upon in this book.

# 1 Introduction
## *History of Life Cycle Costing, Its Categorization, and Its Basic Framework*

*Kerstin Lichtenvort, Gerald Rebitzer, Gjalt Huppes, Andreas Ciroth, Stefan Seuring, Wulf-Peter Schmidt, Edeltraud Günther, Holger Hoppe, Thomas Swarr, and David Hunkeler*

**Summary**

This chapter introduces the historic development of life cycle costing (LCC), dating to the 1930s, and the establishment of the method, based in part on the needs of the military. LCC is defined as a precise tool with 3 variants. Conventional LCC, the historic method, is contrasted with environmental LCC, which uses equivalent systems boundaries and the same functional unit as life cycle assessment (LCA) and which comprises a complementary impact assessment. This new environmental LCC is the main focus of this book. Societal LCC, the 3rd variant, includes an expanded scope still under development, monetizing externalities. This approach stands in contrast to conventional and environmental LCC, which account only for real monetary flows borne by 1 or more actors in the life cycle, even though also anticipated monetary flows in the case of environmental LCC. The idealized case of a washing machine offers examples of considerations for each variant, and the case continues throughout the book as a cross-cutting element in each chapter. The introduction is completed by discussing the key limitations of LCC as preexisting, which should be overcome by environmental LCC and by outlining a general framework for environmental LCC beneficial for a future code of practice or standardization.

## 1.1 HISTORY AND DEVELOPMENT OF CONVENTIONAL LCC

Conventional life cycle costing (LCC), as it is herein termed, is a well-established technique. As early as 1933, in the United States, life cycle costs were included in operating and maintenance costs when the General Accounting Office (GAO) bought tractors. Interestingly, more than 70 years ago, it was understood that acquisition costs were overemphasized in relation to operating and follow-up costs in the procurement process. Indeed, the operating and follow-up costs easily could be several times higher than the initial investment. In the 1970s, LCC was legally mandated for weapon systems procurement by the US government and for the building programs at public institutions in several US states (Society of Automotive Engineers [SAE] 1992). In Europe, LCC has attracted attention in the public sector since the mid-1970s. Increasingly, follow-up costs are allowed for or prescribed in public projects and in building and procurement activities.

Conventional LCC, long before the emergence and development of sustainable development and (environmental) life cycle thinking, was first used in the 1960s by the US Department of Defense for the acquisition of high-cost military equipment such as planes and tanks (Sherif and Kolarik 1981). The rationale was that the purchasing decisions should not be based solely on the initial acquisition cost, but also on the costs for operation and maintenance and, to a lesser degree, for disposal.

Another rationale for conventional LCC is related to optimal budget allocation over the life cycle of a system, product, or item and optimal business performance. Building on this tradition, LCC has so far been applied mainly to decisions involving the acquisition of capital equipment or long-lasting products with high investment costs per unit. Early areas of application have been (Sherif and Kolarik 1981)

- buildings, mainly for commercial or public purposes;
- energy generation and use;
- transport vehicles with high investment costs (mainly from the aerospace sector); and
- major military equipment and weapon systems (see also above).

In the US Department of Defense, several directives for the calculation of life cycle costs and design to costs, both considering already in the R&D phase the cost of a product or a system over its entire life cycle, were developed in the early 1970s (DoD 1973). Also in the United States, several regulations have been issued that require the calculation of life cycle costs for the acquisition of public buildings (Zehbold 1996). However, LCC has usually been limited to sector- or product-specific applications. Sherif and Kolarik (1981) provide a comprehensive overview of the aforementioned applications, the costing models used, and the corresponding literature. They observe that "LCC has developed more as a result of specific applications rather than hypothetical models." This conclusion remains essentially valid today. A generally usable methodological framework or model has not, however, evolved, even though there have been tendencies in this direction (Rebitzer 2005).

The concept that most closely resembles a general LCC method was developed initially by Blanchard (1978) and later refined by Blanchard and Fabrycky (1998); other examples are given in the standards ISO 15663 (International Standards Organization 2000–2001), IEC 60300-3-3 (International Electrotechnical Commission

2004), and AS/NZS 4536 (Standards Australia and Standards New Zealand 1999). These approaches have their roots in systems engineering and focus on the assessment and comparison of technological alternatives. They structure the life cycle of a product or system into research and development, production and construction, operation and support, and retirement and disposal. This structure, in addition to the cost categories considered and the systems view, is very similar to the life cycle approach in life cycle management (LCM) and thus a good basis for the development of an LCC method for LCM. However, Blanchard and Fabrycky (1998) as well as the standards do not elaborate a methodology that gives guidance on how to calculate and compare costs; rather, they present LCC in the sense of "life cycle thinking" and stress the importance of the systems view.

Concluding this short discussion on the historic roots and developments of LCC, one can say that conventional LCC has never been explicitly developed into a broad and generally applicable methodology. Instead it has been developed, based on the principal life cycle view, with the perspective of application-specific procedures in certain sectors. In this context, one can raise the question of why this intriguing and simple concept was never broadly established in industry and the public sector, as for instance quality management approaches were. Recently, some industrial sectors such as the railway industry have started to increasingly acknowledge the importance of LCC for purchase and maintenance decisions. One reason for not applying

**New Challenges beyond Conventional LCC**

Due to the increasing focus on systems thinking, LCC likely will gain more attention over the coming years, particularly for planning and controlling processes and as 1 of the pillars in sustainability analysis, along with LCA and societal assessments.

Existing conventional LCC approaches, or cost management practices, are often not suitable for an assessment of the economic implications of a product life cycle in a consistent sustainability framework. For example, such systems may be incomplete, as when EoL management is missing and environmental effects or impacts cannot be linked consistently to the same system specification as is used for environmental analysis. LCC has to ensure that all the costs of a product or system incurred over its entire life cycle are integrated into the decision-making process to make decisions transparent and to avoid environmental damage and social drawbacks at an early stage. This leads to the conclusion that conventional LCC approaches need to address the complete life cycle and need to be expanded to better link to other sustainability aspects: environmental and, if desired, social aspects. The main question is how costs and environmental aspects can be combined in a consistent way.

This book attempts to address this issue and to provide unambiguous guidelines and examples for 3 distinct types of LCC: conventional, environmental, and societal. Their main features and differences are condensed in this introduction, whereas detailed discussions of the methodologies are presented in Chapters 2 to 6, illustrated by idealized case study boxes, and completed by 7 real LCC case studies in Chapter 7, covering all 3 types of LCC.

a standard LCC method in common business applications is that it frequently would not match the cost system of the company or governmental organization that motivates the LCC study, as a buyer of railways, for example. The LCC method needs to match cost figures used in a company. Any broader scope would be less efficient, and any attempt to provide a more generally applicable LCC method involves to some degree the need to translate company-specific cost figures to more general ones.

## 1.2 TYPES OF LCC

The SETAC-Europe Working Group on Life Cycle Costing has defined LCC (based on the analysis of survey results presented in Chapter 6) according to 3 types. These types are summarized in Figure 1.1, categorized in Table 1.1, and defined, independently, in the following subsections:

*Conventional LCC*: The assessment of all costs associated with the life cycle of a product that are directly covered by the main producer or user in the product life cycle. The assessment is focused on real, internal costs, sometimes even without EoL or use costs if these are borne by others. A conventional LCC usually is not accompanied by separate LCA results. The perspective is mostly that of 1 market actor, the manufacturer or the user or consumer.

*Environmental LCC*: The assessment of all costs associated with the life cycle of a product that are directly covered by 1 or more of the actors in the product life cycle (supplier, manufacturer, user or consumer, and/or EoL actor), with the inclusion of externalities* that are anticipated to be internalized in the decision-relevant future (definition as suggested by Rebitzer and Hunkeler 2003).

Environmental LCC enhances conventional LCC by requiring, on the one hand, the inclusion of all life cycle stages and to-be-internalized costs in the decision-relevant future (hence, anticipated costs), and, on the other hand, separate not-monetized LCA results. A product system according to ISO 14040/44 (2006) should be used as a basis for both. The perspective is that of 1 or more given market actors, mostly manufacturers. If relevant, subsidies and taxes are included in environmental LCC.

*Societal LCC*: The assessment of all costs associated with the life cycle of a product that are covered by anyone in the society, whether today or in the long-term future. Societal LCC includes all of environmental LCC plus additional assessment of further external costs, usually in monetary terms (for example, based on willingness-to-pay methods). The perspective is from society overall, nationally and internationally, including governments. Compared to environmental LCC, subsidies and taxes have no net cost effect, and hence are not included in societal LCC.

All 3 types of LCC have a function-oriented systems perspective, implying a life cycle approach of some sort. The differences between conventional LCC,

* One could define "externalities" in terms of their cost — or, alternatively, either the costs not accounted for in the system or the costs not directly borne by a specific firm. Herein, the 1st definition is used and preferred.

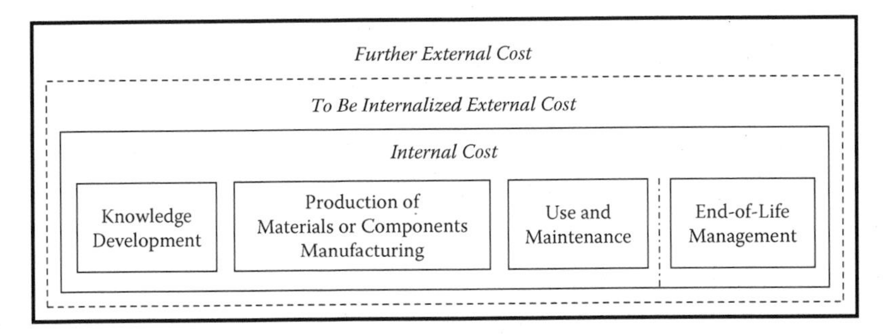


— Conventional LCC: Assessment of internal costs, mostly without EoL costs; no LCA
--- Environmental LCC: Additional assessment of external costs anticipated to be internalized in the decision relevant future; plus LCA in societal = natural boundaries
**—** Societal LCC: Additional assessment of further external costs

**FIGURE 1.1** The 3 types of LCC.

**TABLE 1.1**
**Coverage of the 3 LCC types**

| Aspect | Conventional LCC | Environmental LCC | Societal LCC |
|---|---|---|---|
| Value added compared to conventional LCC | — | Consistent environmental assessment (LCA) at the same time and consistent approach for sustainability assessments of products | Opportunity costs or credits considered |
| Product system (model) | Life cycle, without EoL phase (and sometimes use phase) if not borne by main actor | Complete life cycle | Complete life cycle |
| System boundaries | Only internal costs | Internal costs, plus external costs expected to be internalized | Internal plus all (costs of) externalities |
| Perspectives: actors | Mainly 1 actor, either the manufacturer or the user or consumer | One or more actors connected to the product life cycle, mainly manufacturers, supply chain, and end user or consumer | Society overall, including governments |
| Reference unit | Item or product | Functional unit | System |
| Cost categories | Mainly acquisition costs (research and development [R&D] costs and investment costs) and ownership costs (operating costs, maybe disposal costs) | Mainly costs of development, materials, energy, machines, labor, waste management, emission controls, transport, maintenance and repair, liability, taxes, and subsidies | Mainly costs of construction, maintenance, and environmental damages; taxes and subsidies have no net cost effect |

*(continued)*

**TABLE 1.1**
**Coverage of the 3 LCC types (continued)**

| Aspect | Conventional LCC | Environmental LCC | Societal LCC |
|---|---|---|---|
| Cost model | Generally quasi-dynamic model | Steady-state model | Generally quasi-dynamic model |
| Discounting of result of LCC | Recommended (but usually not applied) | Inconsistent and not recommended | Recommended |
| Discounting of cash flows for calculation | Recommended | Recommended | Recommended |
| LCA according to ISO14040/44 (2006) | No | Yes | Not recommended due to risk of double counting and inconsistencies |
| Standards and guidances | Various (ISO, IEC, SAE, AS/NZS, etc.) | None (LCA: ISO 14040/44 2006) | For various elements thereof, including from the United Nations (UN) and Organization for Economic Cooperation and Development (OECD; Dasgupta et al. 1972; Little and Mirrlees 1969) |
| Use in life cycle management (LCM) | Mostly internal decision making to private organization and supply chain considerations | Mostly internal decision making of producer or user of product, but also for external communication (similar to LCA) | Mostly internal to public organizations |

environmental LCC, and societal LCC are, as mentioned, categorized in Table 1.1. This book focuses on the new methodology of environmental LCC, whereas conventional LCC is well known (see Section 1.2.1) and societal LCC still suffers from not being completely defined by the scientific community (see Section 1.2.3).

### 1.2.1 Conventional Life Cycle Costing

Even though conventional LCC is demand driven, leading to separate application fields, several standards regarding harmonizing conventional LCC application are available: for example, IEC 60300-3-3 (International Electrotechnical Commission 2004), ISO 15663 (International Standards Organization 2000–2001), SAE-ARP4293/94, DoD 1973 (US Department of Defense 1973), and AS/NZS 4536 (Standards Australia and Standards New Zealand 1999). In addition, different methods for performing conventional LCC have been described in the literature (e.g., Dhillon 1989; Ellram 1993, 1994, 1995; Fuller and Petersen 1996; Riezler

1996; Zehbold 1996; Australian Department of Defence 1998). This well-known knowledge will not be repeated in this book; rather, the differences to environmental LCC will be demonstrated.

Two important traditional approaches to costing are closely related to LCC, namely, total cost of ownership (TCO) and activity-based costing (ABC). TCO helps consumers and enterprise managers to assess the total costs related to the use of an item (Ellram 1993). TCO and users' or consumers' LCC do overlap, as TCO has a strict user perspective focusing in particular on the acquisition and use phase (investment, maintenance, operation, support, etc.). ABC supports manufacturers to calculate the true costs of an item by assigning overhead and other general costs to products (including services), in addition to direct and indirect costs (see Roztochi 1998). However, in both approaches, no environmental assessment or external costs are included, and ABC generally lacks the life cycle perspective, hence it does not qualify as LCC (see Chapter 3).

### 1.2.2 Environmental Life Cycle Costing

An argument for environmental LCC, albeit a subtle one, is that assessment methods such as LCA are often viewed as obstacles to business development, particularly in the short term (Hunkeler and Rebitzer 2003). A new methodology that provides a sound combination of both the environmental and economic performance of a product can help with guiding technological development and managerial decisions in a more rational direction, identifying win–win situations, and optimizing trade-offs between the environmental view and the economic and business view.

The general scope of environmental LCC is defined within a conceptual framework outlined in Figure 1.2, with which the relationship of LCC to LCA and societal assessments (e.g., including employment conditions, unemployment rates, and general social impacts on communities) in LCM can be explained.

The conceptual framework of environmental LCC is based on the physical product life cycle, discerning 5 stages that can be further refined if needed: research and development, production of materials or components, manufacturing, use and maintenance, and EoL management.

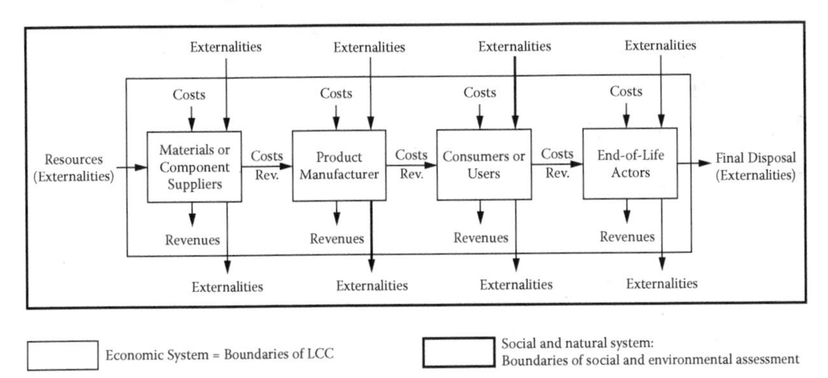


**FIGURE 1.2** Conceptual framework of environmental LCC. *Source:* Rebitzer and Hunkeler (2003).

In Figure 1.2, the actors of the 5 life cycle stages are represented: material or component suppliers dealing with research and development (R&D) and production of materials or components, product manufacturers dealing with R&D and manufacturing, consumers who use and maintain products, and EoL actors dealing with EoL management.

One can differentiate between 2 kinds of costs (Rebitzer and Hunkeler 2003):

- Internal costs along the life cycle of a product, with "internal" implying that someone — a producer, transporter, consumer, or other directly involved stakeholder — is paying for the production, use, or EoL expenses, and, thereby, the internal costs can be connected to a business cost. This cost definition concerns all the costs and revenues within the economic system (inside the fine lines in Figure 1.2). Internal costs can be divided into costs inside or outside an organization, depending on the perspective.
- External costs (externalities) that already are priced in monetary units, due to their to-be-internalized character in the decision-relevant future, and remain so; there is no conversion from environmental measures to monetary measures, or vice versa. There should be no double counting of externalities in LCC and the complementary LCA.

It may be noted that cash in–cash out instead of costs–revenues could be distinguished in Figure 1.2, with opposite flow directions as shown. However, this concept is not recommended for environmental LCC because this type of LCC is based on the same concept as LCA with its defined material and energy flow direction. Environmental LCC will be presented in detail in Chapter 3.

### 1.2.3 Societal Life Cycle Costing

Cost–benefit analysis (CBA) may be considered as a source of ideas regarding how to take a social cost perspective into account for the development of societal LCC. CBA has been developed for major public investment plans and looks customarily for costs, and benefits, over the lifetime of the project or program investigated. The 1st cost–benefit analyses were set up in relation to large public works like the Tennessee Valley reconstruction in the United States in the 1940s and the Delta plan for total sea dyke renovation in the Netherlands after the flood disaster of 1953 (Tinbergen 1961). Since then, the economic methods have developed increasingly toward quantifying intangibles in monetary terms and integrating the outcomes of a CBA in 1 or a few figures. Because all actors in society are included, the payments between them cancel out, and only the value added in all activities involved remains as cost. For further details on CBA, refer to Dasgupta and Pearce (1972), Pearce (1983), or Mishan (1975).

Societal LCC assesses all costs associated with the product life cycle covered by anyone in the society. In a societal LCC, not only "real money flows" are considered but also externalities (i.e., value changes caused by a business transaction but not included yet in its price or benefit). Environmental LCC may already consider the monetization of externalities in the decision-relevant future (e.g., upcoming costs for $CO_2$ emission-trading certificates or global warming adaptation costs). Societal LCC goes beyond this concept by considering any externality that could be monetized, for example damage costs evaluated in the ExternE Project (Bickel and Friedrich 2005), or even those that are difficult to monetize and may therefore only be considered qualitatively, for example public health and social well-being, standards and rights related to job quality, and private and family life. The limits are obvious because, on the one hand, there are many externalities that may be monetized (see Chapter 4 and Table A.2, "Appendix to Chapter 4: Social Impacts"), and, on the other hand, an externality can only be considered if it can be detected and if an actor in the product life cycle knows that he or she is affected and cares.

The brief analysis above reveals that societal LCC is used to quantify environmental effects on society in money terms and may be seen as a useful concept for linking environmental life cycle approaches to corporate social responsibility or policy decisions. This connection is a matter of taste and perspective, however. The advantage of having a single score in welfare terms has to be weighed against the high uncertainty on the evaluation of social effects, if really known, as well as practicality and transparency (see the goals of LCM in Hunkeler et al. 2004). It may then be more acceptable to leave the social impact assessment score separate, to be compared to other costs (for a discussion on the limits of such a monetization of externalities, see Ackerman 2004). The latter would be in line with the Brundtland definition of sustainable development (Brundtland Commission 1987), which means that implications in sustainability can only be analyzed and balanced (trade-offs and/or win–win situations) if the social, environmental, and economic dimensions are left separate. It is therefore recommended to present disaggregated results, and not a single figure, for the different impact categories and to carry out sensitivity analyses (as is also the case for LCA by itself). This recommendation would enable a governmental organization, as a main target group for societal LCC, to consider all benefits and costs to a society that are caused by a decision, legislation, or strategy. For example, the European Commission is following this 3-pillar concept in its policy impact assessments, which are separated into disaggregated economic, environmental, and social impact assessments (European Union 2005b).

## 1.3 TWO KEY LIMITATIONS OF LCC TO BE TACKLED BY ENVIRONMENTAL LCC

The need for different perspectives, depending on the actor in the life cycle (manufacturer, user, etc.), and the issues of prospective LCC versus retrospective LCC are heavily debated topics in LCC that are not solved in conventional LCC so far. These limitations, discussed in Sections 1.3.1 and 1.3.2, should be overcome by environmental LCC, which is presented in detail in Chapter 3 as the core innovative type of LCC in this book.

### 1.3.1 Need for Different Perspectives

To date, the life cycle approaches defined in the literature are limited seemingly because the life cycle is managed as an entity in itself. It is not specified who decides on the single production steps or defines the user demands, the fulfillment of which is, after all, the ultimate aim of every good or service offered. The life cycle of a product forms a given issue, which seems outside the control of the single companies taking part in it. Therefore, 1 open issue is the question of whose costs one is accounting for (or managing). Specifically, are the costs of the user or consumer, the producer, or the waste management operator, for example, the relevant ones? This question is caused by the existence of value added, which has no counterparts in LCA (Rebitzer and Hunkeler 2003). Because the cost for 1 actor (e.g., the consumer who buys a product) is the revenue for another, the cost for actors in the chain cannot be summed. Only the full system perspective gives a relevant total comparable to the total environmental impacts. Ideally, one would seek to use the same basic, full LCC analysis and derive the costing from the perspective of various stakeholders for the same underlying data.

The answer to the aforementioned question depends on the objectives of the LCC study resulting from the perspectives of the actors. For example, in a wastewater treatment LCC (Section 7.2), the costs in the focus are those of the waste treatment, all other costs being "sunk." In a consumer product LCC (cradle-to-grave perspective), all costs are included, whereas in a producer's LCC, costs up to the product's arrival at the gate are considered plus implications for market success due to use and disposal costs. In more general terms, a full LCC can be split up into costs relevant for a specific decision maker within that system. However, the direct and indirect feedbacks relevant for 1 actor (e.g., costs for a consumer, which influence the purchasing decision) should be considered.

The decision of the target group for an LCC also has implications on the necessary level of detail. If the perspective of the assessment is that of the user or consumer (Figure 1.3, part c), the costs within the boundaries of the other organizations or actors can be viewed as a total only (black box), without requiring any differentiation. Of particular interest, however, are the specific costs and revenues associated with the use of the product (e.g., introductory and energy costs, maintenance, and reselling of the product). On the other hand, if a manufacturer seeks to optimize the life cycle costs, the detailed process costs that can be allocated to a product within the company are the major focus (e.g., also using activity-based costing to allocate overheads). In this case, the other cost categories in the life cycle require less detail (Figure 1.3, part A).

Part B in Figure 1.3 represents a case in which the level of detail within different actors or organizations is important. This is the situation if, for example, the supply chain is integrated by the acquisition of the supplier or by supply chain coordination efforts (Seuring 2002). The notion of value added requires one to consider both costs and revenues in each stage for LCC (Rebitzer and Hunkeler 2003). Value added is the difference between the sales of products and the purchases of products or materials by a firm, covering its labor costs and capital costs as well as its profits. These elements together form the value added of the firm or activity group. The value added of the product manufacturer is the difference between the revenues from sales to the consumer or user and the costs of purchases from the suppliers of materials or components, disregarding possible revenues from co-products (by-products).

Environmental LCC provides results for all different perspectives presented in Figure 1.3 (i.e., for product manufacturer, supplier, user, or end-of-life actor; see also

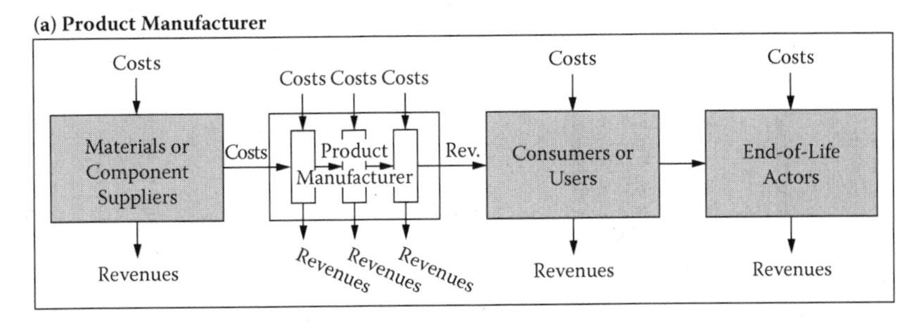


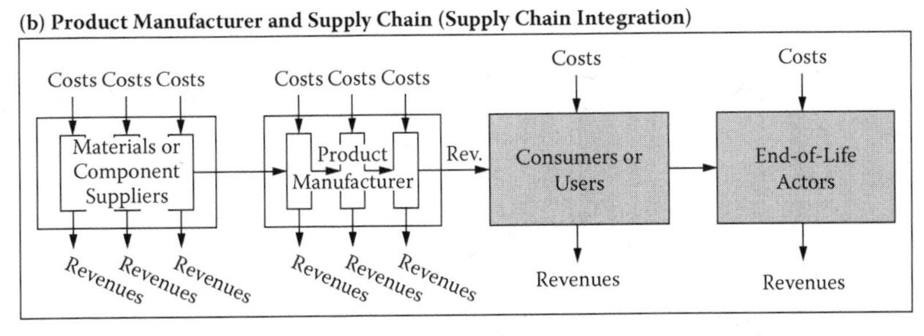


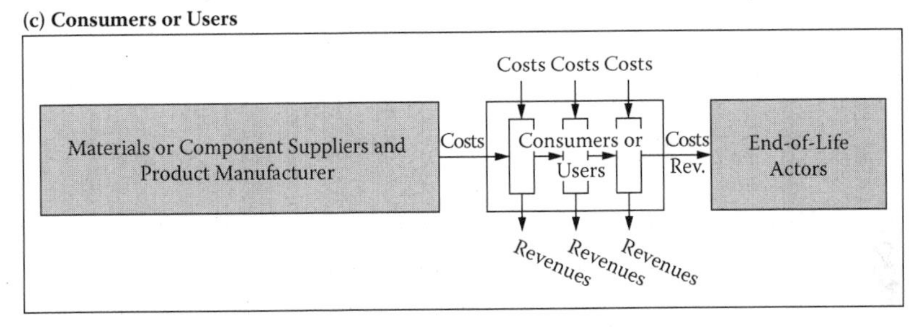


**FIGURE 1.3** Different perspectives in LCC (nonexhaustive examples). *Source*: Rebitzer and Hunkeler (2003).

Case Study Box 3: Perspectives in Chapter 2). This advanced result presentation is consistent with the goal and scope definition as presented in Case Study Box 1: Goal and Scope Definitions from Different Perspectives (below).

### 1.3.2 Life Cycle Costing Planning versus Life Cycle Costing Analysis

The LCC process should also be differentiated with respect to when in the life cycle it is carried out (see Figure 1.4).

LCC planning is performed during the design or planning stage of the product life cycle and is used to compare and assess options and alternatives (prospective, consequential, or change-oriented LCC). When prospective LCC is linked to LCA results, they can support the integration of environmental considerations into product development (see ISO 14062 2002; or Schmidt 2003).

LCC analysis deals with existing products and is applied to monitoring and managing costs over the product's life cycle. Also, LCC analysis may refer to past product systems to gather information for future products.

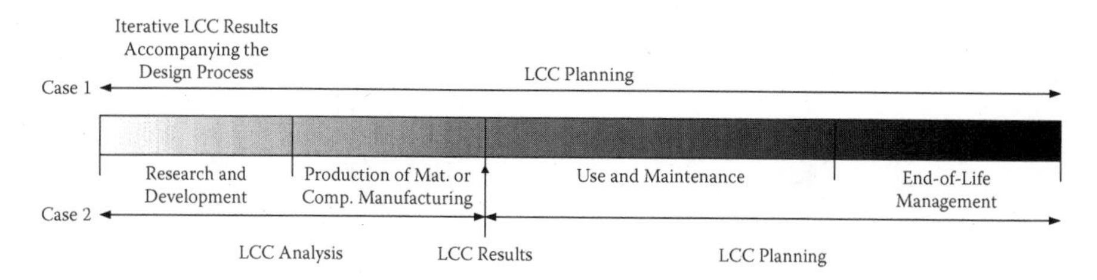


**FIGURE 1.4** LCC planning versus LCC analysis.

So we may have prospective LCC or retrospective LCC. Environmental LCC, due to its focus on decision support, favors a future-oriented LCC approach — taking into account that most of the costs and environmental impacts are being fixed during the design phase — which corresponds rather to the LCC planning approach (see case 1 in Figure 1.4). However, depending on the study's goal and scope, the LCC analysis approach will also be adequate if accounting for and reporting of (historical) cost data and related information comprise only 1 part of the ultimate goal, with another aim to support future-oriented management decisions (see case 2 in Figure 1.4). It is obvious that LCC planning has to deal with higher uncertainties than LCC analysis since it does not relate to already existing products.

## 1.4 THE REQUIREMENT AND GENERAL FRAMEWORK FOR ENVIRONMENTAL LIFE CYCLE COSTING

As mentioned at the outset of this chapter, environmental LCC is not envisioned as a stand-alone technique, but is seen as a complementary analysis to environmental LCA. Therefore, environmental LCC should be developed in analogy to LCA, which means that this book serves as a precursor to a code of practice for environmental LCC, which should define a general framework for environmental LCC that is beneficial for a potential standardization in analogy to ISO 14040/44 (2006). At the current stage of development, the following steps may well be relevant for carrying out consistent environmental LCC, although the specific steps in the analysis can vary from case to case:

1) Goal and scope definition
2) Information gathering
3) Interpretation and identification of hotspots
4) Sensitivity analysis and discussion

These steps are briefly described in the following subsections.

### 1.4.1 Goal and Scope Definition

The goal and scope of environmental LCC need to be defined before a study takes place. It is critical to appropriately define the system boundary as well as the functional unit. The following possible additional objectives for LCC have been identified by surveying different expert opinions:

- Identifying total costs for the actor itself (e.g., a firm or consumer)
- Assessing competitiveness of product (identifying consumer costs, i.e., the cost of ownership)
- Reporting, monitoring, and proactively influencing costs within companies
- Achieving agreement at a management level on product portfolio development and selection and on how LCC is linked to the product portfolio
- Identifying possible alternatives for development or marketing
- Identifying economic and environmental trade-offs and win–win situations
- Addressing corporate social responsibility (CSR), specifically if carried out together with societal assessments
- Identifying a potential business case and examining long-term costs, examining potential economic benefits for consumers or at the EoL stage, and flagging potential economic, environmental, or societal risks
- Defining trade-offs between various criteria, for example, future versus current costs, or internal versus external costs
- Defining life cycle optimization questions such as the change of maintenance regimes for a purchased product

As an example, typical goal and scope definitions for a washing machine from different perspectives are given in Case Study Box 1.

LCC, like LCA, usually compares alternatives: basic alternatives, as between a train and airplane for the function of transportation, or those arising from differences in activities that occur in specific processes in an otherwise identical life cycle. In the latter case, there is simplified data collection because only cost differences need to be specified.

It is important to note that the selection of alternatives must be consistent with the functional unit as defined in LCA (ISO 14040/44 2006). An example is the comparison of beverage bottles, in which disposable plastic or reusable glass bottles may be compared. The functional unit should be a given utility resulting in different reference flows (e.g., 1 reusable 1 L glass bottle that is used 20 times versus 20 disposable 1 L plastic bottles that are used once). Peer review of the definition of the functional unit, system boundary, and list of options, which all (among others) should be defined in the goal and scope definition, seems both prudent and necessary.

### 1.4.2 Information Gathering

If all needed data are not available, then scenario development, forecasting, or other estimation methods may have to be employed (as in LCA; see Rebitzer and Ekvall 2004). Fortunately, the field of cost estimation is very well established (Dhillon 1989). Furthermore, thresholds can be applied, as is the case in LCA. Here, typically, a process-specific threshold related to mass of input materials below 5% could be justified to be neglected (i.e., applied as a cutoff) as it would usually not influence the LCA result (Rebitzer 2005). For LCC, such thresholds remain to be benchmarked, though in a given life cycle stage (e.g., production, transport, or use) a quite conservative rule of thumb could be 1% of the total cost of that stage. However, if this cost estimate has been made, it may be included as a lump sum that is not further detailed.

### Case Study Box 1: Goal and Scope Definitions from Different Perspectives

For the illustration in the case study boxes, a washing machine has been selected. The washing machine is an investment good purchased by households with a lifetime between 10 and 15 years. There could be different goal and scope definitions depending on the perspectives (e.g., that of the manufacturer, the supplier, the EoL actor, the consumer, and society).

#### Manufacturer's Perspective

*Option I*

Because the responsibility for the EoL lies with the manufacturer (assuming that a corresponding legislation applies), different scenarios of the EoL will be compared in an LCC study. EoL scenarios include the collection of the used washing machine at the household (or via other collection systems), transports to recycling companies, and different scenarios for recycling technologies and market demand for secondary materials produced. Future market and recycling infrastructure conditions are considered in scenarios with varying prices for primary and secondary resources as well as EoL operating costs.

*Option II*

Another goal and scope definition of a manufacturer targets the comparison of various scenarios for manufacturing technologies.

#### Manufacturer, Supplier, and End-of-Life Service Provider

Together with the supplier, who delivers a component that causes significant effort at the end of the life of a washing machine, the manufacturer and the EoL service provider compare alternatives to reduce costs and environmental impacts.

#### Consumer's Perspective

The goal of the LCC is to investigate whether it makes sense to further use an existing washing machine in stock or to replace it with a new, more efficient one before its actual breakdown. The question shall be answered from an environmental and an economic perspective. To this end, the further use of washing machines of different ages (i.e., only the use phase is accounted for) is compared to the acquisition and use of a new washing machine (including the replacement of the old one, i.e., the production or acquisition, use, and EoL phases are considered). Thus it can be determined after what time period the initial environmental impact as well as costs through production and acquisition (and recycling) are compensated by the lower impacts and costs, respectively, during the use phase through the use of the more efficient new appliance (i.e., the payback period). Changing market conditions in the future are considered in scenarios with varying market prices for resources like energy and water and costs for sewage treatment.

*Note*: These are the goal and scope of the washing machine case study presented in Section 7.5.

**Societal Perspective**

The washing machine LCC study shall provide information about the environmental savings potential and the related holistic costs that result from different improvement strategies. The most promising strategy shall then be communicated to private consumers via a publicly funded information campaign. In the case at hand, there are basically 2 strategies, which are compared to the base case (average low-cost washing machine, and average behavior): 1) buying a more efficient washing machine or 2) optimizing the user behavior (e.g., higher loading and lower washing temperatures). The study compares all costs that have to be paid by the consumer during the whole life cycle of washing machines for the considered alternatives. The lifetime begins with the purchase of the washing machine and assumes an average life span of 1840 washing cycles. Costs that are considered include acquisition costs and costs for energy, water (including sewage), and detergent use.

*Note*: These are the goal and scope of the washing machine study according to Rüdenauer and Grießhammer (2004)

One problem in data gathering is that of allocating costs to products that are produced together with co-products or by-products. In LCC, the costs of personnel and capital and of acquired goods and services are to be allocated to the different outputs involved, based on their market prices. Cost allocation methods generally are of the gross sales value method, in which the costs are allocated specifically back to a "split-off point," with all upstream cost based on the share in (thus, adjusted) proceeds (Huppes 1993).

### 1.4.3 Interpretation and Identification of Hotspots

A key outcome of an LCC, as well as of an LCA, is the identification of hotspots. These hotspots usually become evident as a result of the analysis, particularly if a sensitivity analysis (Section 1.4.4) is carried out. Interpretation of these hotspots can be quantitative or qualitative. For quantitative analysis, classical methods of investment appraisal such as net present value, annuities, internal rate of return, and payback period can be applied (see Chapter 2 for the related, more detailed discussion).

Assessments of alternative systems are often influenced by nonmonetary (possibly qualitative) criteria. An alternative that might be optimal from a quantitative point of view can be rejected based on other aspects analyzed (e.g., not having the required market appeal and thus not providing the required or desired sales volume). Absolutely advantageous alternatives occur only if their performance is superior in both the quantitative LCC and other, possibly qualitative, analysis. In the case of diverging scores, the decision maker must assess each alternative individually. Several methods for decision support combining quantitative and qualitative aspects have been elaborated for LCM; therefore, they are not elaborated here (see Hunkeler et al. 2004).

### 1.4.4 Sensitivity Analysis and Discussion

Sensitivity analyses are not, per se, required for LCA, though they are recommended in the interpretation phase, and they are not generally used in conventional LCC. One would hope, and perhaps expect, that an ultimate LCC code of practice or standard would, however, mandate sensitivity studies for environmental and societal LCC.

Connections between uncertain parameters used in LCC (e.g., project life, included life cycle costs and revenues, sales volume, and/or the discount rates) and calculated outputs (e.g., net present value) should be revealed by a sensitivity analysis. The key question is how sensitive the outputs are to given deviations in the input parameters. To elaborate this, the uncertain input parameters are varied ceteris paribus by a certain percentage and their effects on the output parameters are noted. Thus, it is possible to determine how outputs vary for given variations in inputs. Sensitivity analysis also provides an answer to the question "To what extent can the input values vary without impacting the conclusions related to comparing different options or causing the output values to deviate from a certain value?" A disadvantage of sensitivity analysis is that only 1 input can be varied at a time. Monte Carlo simulation resolves that problem, though it is much more complex and less transparent. The sensitivity analysis should be the basis for the final discussion and the development of recommendations.

# 2 Modeling for Life Cycle Costing

*Gjalt Huppes, Andreas Ciroth, Kerstin Lichtenvort, Gerald Rebitzer, Wulf-Peter Schmidt, and Stefan Seuring*

**Summary**

This chapter discusses the time value of money as well as how discounting should be carried out so that the estimated life cycle cost is consistent with the methodology employed. Discounting will depend on the type of life cycle costing (LCC) carried out as well as the dominant environmental impacts, and is an iterative procedure requiring a sensitivity analysis and peer review. The need to consider LCC from the perspective of who bears the cost is highlighted in a case study. Explanations are given as to when it is appropriate to include taxes, tariffs, and externalities such as willingness-to-pay values. The aggregation of costs is also summarized.

## 2.1 INTRODUCTION

One could certainly question how fundamental the differences are between the types of LCC and what practical consequence these variations have in carrying out the analysis. Within this chapter, the dimensions of costing are examined, each one attempting to respond to a set of questions that may arise when one is involved in collecting, or estimating, the costs to be included in an LCC, including the following:

- *How are costs modeled?* Are the costs reported, evaluated, and distinguished over time, as with (quasi-)dynamic modeling, or is the time value of money not considered?
- *Which cost categories are employed?* Are only market costs considered, or does the analysis expand to include taxes and tariffs or even concepts such as willingness to pay?
- *Whose costs are taken into account?* Are only the costs from specific firms and individuals considered, or are costs from the society at large included?
- *How are costs aggregated?* Are costs reported as averages, in terms of net present value, or as annuities?

Each of the aforementioned questions relates to 1 of the 4 basic dimensions of LCC and will be elaborated upon in the following sections. Throughout this chapter, case study boxes based on real, and partly hypothetical, washing machine LCC are used to demonstrate the outcomes of different methodological choices.

## 2.2 COST MODELS

Cost modeling is characterized by how the time value of money is considered and the degree of nonlinearity relating outputs to inputs. For example, the LCA model, as a whole, is linear homogeneous or homogeneous to degree 1, implying that twice the input produces twice the output (as is the case for mass and energy balances in general, as long as no nuclear reactions are involved). In economic theory this relation is typified as "constant returns to scale." In sophisticated cost modeling, neither of these characteristics is justified and required. First, there are modeling types that use exponential relations and still are linear homogeneous, such as Cobb-Douglas production functions. Second, models may use linear relations but do not exhibit constant returns to scale, like most optimization models. Furthermore, the majority of most nonlinear relations will also lead to nonlinear homogeneous models, with increasing or decreasing returns to scale or more complicated relations. The characteristics of the different models that can be employed in LCC are summarized in the following discussion.

Steady-state models are conceptually the simplest ones, owing to the fact that they lack any temporal specification and assume all technologies remain constant in time. Most LCA applications are steady-state models, as are substance flow analysis (SFA) and input–output analysis (IOA) models. This is the approach employed in environmental LCC (Huppes et al. 2004).

Quasi-dynamic models are time series that are exogenously determined. They are a compromise between steady-state and dynamic models. These models assume that most of the variables remain constant in time, though they allow one or more of them to vary. Most CBA and some IOA models are quasi-dynamic. Conventional and societal LCC are, generally, quasi-dynamic.

Dynamic models explain the development of variables over time, with past values determining future ones. For example, economic models may predict investments in the following year based on the profits of this year. In contrast to quasi-dynamic models, these values are derived endogenously. Macroeconomic models often are dynamic models.

For conventional and societal LCC, the use of quasi-dynamic models makes it difficult to directly compare the results with steady-state environmental methods (i.e., LCA). Therefore, environmental LCC is primarily set up as a steady-state method, designed to be compatible with LCA. Some aspects of societal assessment, the 3rd pillar of sustainability, may be linked to the steady-state type of modeling as well, though highly relevant items such as income distribution and unemployment rates have a dynamic background. A clear disadvantage of the steady-state approach to LCC is for firms in that the quasi-dynamic approach (i.e., conventional LCC) is the relevant way of comparing the cost of options or the attractiveness of investments. However, surveys indicate (see Chapter 6) that some corporations are coupling steady-state environmental assessments and quasi-dynamic LCC.

**Summary: Temporal Modeling Is a Key Parameter in LCC**

Effectively, the modeling choice is between steady-state models, linked to environmental LCC and quasi-dynamic models, consistent with conventional and societal LCC. As life cycle assessment is steady state in nature, environmental LCC is the most compatible of the 3 methods to be employed in sustainability assessment.

## 2.3 COST CATEGORIES

External costs either are market based or resemble other money flows connected to a product's life cycle (e.g., taxes and tariffs). These should be distinguished from the cost of external effects. Such externalities (see Chapter 4) include concepts such as willingness to pay (for avoiding these effects) or the cost of preventing the effects. Though it may seem like a nuance, external costs are part of the product system and should be considered in all types of LCC, while externalities are extremely uncertain to be monetized in the decision-relevant future and are, therefore, only considered in societal LCC.

### 2.3.1 Cost, Revenue, and Benefits

Consider the following example as an illustration. In multifunctional refinery production, LCA has 2 options to deal with product flows coming out of the refinery: to split up the refinery virtually, as by economic (or other) allocation, or to subtract the co-products, as by substitution. In cost terms, the economic allocation has an easy equivalent in cost allocation as applied in managerial accounting (cost management; Rebitzer 2005). The equivalent of LCA-type substitution is subtracting the cost of some other production process having the same output. This method is, at times, applied in national (macroeconomic) accounting, though never in cost management. The substitution equivalent does not exist in LCC. The method applied is that of cost allocation, indicating which part of total cost, including profits, is due to each of the products sold, of course reckoning the cost due to just 1 of the products first. This example illustrates that there are good reasons to explicitly treat both cost and revenues in LCC and to specify how the revenues are dealt with. There seem to be no fundamental problems involved in adding the revenues in the analysis, as long as it is clear how it is being carried out. For very practical reasons, revenues are frequently left out, if they may be assumed to be rather identical for different product systems being compared, or if they are very small as compared to costs.

### 2.3.2 Market Prices and Value Added

In national accounting the national product may be determined based on market prices or factor costs. The total of both is the same, though the means of arriving at the total are quite different: adding all expenditure on products, leaving out all intermediate sales, or adding up all factor costs, as payments for capital and labor. In LCC these approaches may be combined. However, under such circumstances it should be clear which method is employed where. From the point of view of a certain firm — and quite similar situations apply to some public organizations — costs are reflected in the prices paid for products acquired and in the cost for providing capital goods and labor. When comparing the sales of a firm with the costs of products acquired by it, the difference is the gross value added: that is, the sum of labor costs and capital costs, including profits (excluding value-added tax [VAT] and other taxes). This value added may be left gross, or may be made net, after deduction of what is set aside to compensate for the wear and tear of the capital goods (i.e., depreciation). Capital goods acquired hence should not be lumped to other goods acquired, but should be covered by some measure of depreciation. The cost of borrowing (loans, leases, etc.) should be included as well, as should profits, which remain after deduction of the cost of borrowing. The treatment of depreciation and taxes is a delicate subject, as there are many conventions in different countries. Furthermore, conventional LCC often employs direct cash flows (i.e., without depreciation). In national accounting, these difficulties have been resolved, one way or another.

One way to avoid the aforementioned difficulties in LCC is by not detailing cost from a firm's point of view. Each product system, close to its kernel process, has a limited number of products together delivering the service(s) as specified in the functional unit. Taking the (expected) market prices of just these products, including the waste disposal services implied in using the product, would provide the total life cycle cost. This simple method has 1 disadvantage in that it does not give insight regarding which factors determine costs, essentially making sensitivity analysis impossible. Furthermore, if alternative technologies are involved that are not yet on the market, it is not possible to use market prices. Then, more detailed in-firm type cost functions are to be used as models for specifying the cost.

**Summary: Accounting and Financial Definitions**

The cost of purchases, in market prices, reflects the total upstream gross value added. Adding the gross value-added figures of the firm gives the total value of output of the firm, its sales, as the cost of purchases of the next actors in the chain. The gross value added is the sum total of labor cost and capital cost, including depreciation and profits. LCC requires rigorous accounting of cost categories (even if not detailed) and transparent definitions.

### 2.3.3 Four Levels of Cost Categories

Four levels of cost categories may be distinguished: economic cost categories, life cycle stages, activity types, and other cost categories (see Table 2.1). When making an LCC analysis, these 4 levels are best decided on sequentially. In particular, and when applied in a decision-oriented context, the 3rd and 4th levels are most relevant.

**TABLE 2.1**
**Overview of cost categories**

| Level | Cost category | | |
|---|---|---|---|
| 1st level: economic cost categories | Budget cost, market cost, alternative cost, and social cost | | |
| 2nd level: life cycle stages | Knowledge development (including R&D), primary production (materials, energy, etc.), components production, manufacturing, use, and end-of-life management | | |
| 3rd level: activity types | Development, extraction, purchase, sales, reuse, and management | | |
| | Design, agricultural production, schooling, public relations, recycling, and administration | | |
| | Research, testing, packaging, transport, maintenance, waste processing, and infrastructure | | |
| 4th level: other (exemplary) cost categories 1 | Conventional cost | Transfer payment | Environmental cost (internal) |
| | Personnel and equipment costs, rents, and profits | Direct taxes | Damage prev. costs |
| | Materials disposal, communication costs, and investments | Indirect taxes | Wastewater treatment costs |
| | Food production, services, electricity, and office cost | Excises and levies | Exhaust gas reduction costs |
| | Building costs, warranties, infrastructure costs, and depreciation | Subsidies | Rehabilitation costs |
| 4th level: other (exemplary) cost categories 2 | *Management*: material cost, energy cost, personnel cost, machinery cost, transport cost, disposal cost, revenues, and end-of-life value<br>*Supplementary*: service cost, tooling cost, storage cost, taxes, warranties, assurances, infrastructure cost, building cost, settlement cost, control cost, financing cost, and appliance cost | — | Residual value |

This 1st level corresponds roughly with the choice on the family of LCC, as is documented in the following summary box.

**Summary: Relevance of Cost Categories**

Budget cost and market cost are relevant for conventional LCC. Alternative cost and social cost are the prime cost types for societal LCC, whereas transfer payments (taxes and subsidies) are not considered. For environmental LCC, a choice has to be made. In principle, the full systems point of view suggests an alternative costs type, including the net of transfer payments from and to governments. However, for the practical purposes of the majority of business- and consumer-focused analyses, market costs are likely adequate.

The 2nd level has to do with the completeness of the system. In principle, all stages in the life cycle should be included. However, from the point of view of an individual firm, the sum of its internal cost, as value plus its costs of external purchases of products (covering both goods and services, including waste management services), equals its cradle-to-gate cost level and hence does not correspond to full life cycle costs that would generally include use, transport, and end-of-life expenditures.

The 3rd level reflects the life cycle stages in more detail and may especially be useful to track overheads, quite often neglected in LCA systems specification, though possibly coming in view when a hybrid approach is applied, using environmentally extended input–output data (Suh et al. 2004; Suh and Huppes 2005) for background data. The activity types distinguished in Table 2.1 may easily be expanded systematically using the EU nomenclature as developed in NACE (Nomenclature Générale des Activités Économiques dans les Communautés Européennes) and its US equivalent, NAICS (North American Industry Classification System), both involving several hundred well-described activity types. These classifications of activity types have a global origin, being based on the International Standard Industrial Classification (ISIC) classification of the United Nations.*

In the 4th level, the most specific cost categories are distinguished. Case Study Box 2 illustrates the cost categories discussed herein.

* The United Nations Statistics Division (UNSD) has developed a standard product classification as well, as applied in make-and-use tables, the HS (Harmonized System), and has developed a nomenclature for final consumption by private consumers and governments, Classification of Individual Consumption according to Purpose (COICOP). For a related survey, see United Nations Statistics Division (2007). For environmental cost, the European Classification of Environmental Protection Activities and Expenditure (CEPA) can act as a guide.

## Case Study Box 2: Cost Categories

This case study box illustrates the cost categories chosen to calculate an environmental LCC for a washing machine. Budget costs and market costs are considered for all life cycle stages (manufacturing, use, and EoL), whereas the conventional cost categories and some transfer payments are allocated to the actors of each life cycle stage.

For the R&D phase, only the labor costs of the washing machine designers are taken into account. The preproduction phase is considered via all costs for the materials and components necessary to produce the washing machine, whereas production costs such as electricity, gas, water, and so on are added for the production stage.

Private households have to regard the purchase costs for the washing machine and operating costs such as water, electricity, and detergents. In this example, it is assumed that there are no direct end-of-life costs for the consumer due to take-back regulations (disassembly costs minus reuse revenues, or recycling costs minus secondary material revenues).

| | Amount | Cost per unit | Costs |
|---|---|---|---|
| **Appliance Manufacturing** | | | |
| ***Research and Development*** | | | |
| Labor | 0.5 hours | 40 € / hour | 20 € |
| ***Components or raw material production*** | | | |
| Steel | 26.5 kg | 1.5 € / kg | 39.75 € |
| Concrete (weight) | 1 piece | 10 € / piece | 10.00 € |
| Carboran 40% | 12.0 kg | 1.8 € / kg | 21.60 € |
| Plastics (mainly polypropylene [PP]) | 6.0 kg | 1.1 € / kg | 6.60 € |
| Aluminum | 4.0 kg | 1.8 € / kg | 7.20 € |
| Chipboard | 2.5 kg | 0.9 € / kg | 2.25 € |
| Gray cast iron | 2.0 kg | 1.2 € / kg | 2.40 € |
| Glass | 1 piece | 16 € / piece | 16.00 € |
| Copper | 1.0 kg | 1.9 € / kg | 1.90 € |
| Electronic components | 1 piece | 75 € / piece | 75.00 € |
| Cotton with phenolic binder | 0.5 kg | 35.0 € / kg | 17.50 € |
| Cable | 1.5 m | 1.5 € / m | 2.25 € |
| Other materials | 2.0 kg | 7.0 € / kg | 14.00 € |
| Sum | | | 216.45 € |
| ***Production*** | | | |
| Electricity | 50.0 kWh | 0.16 € / kWh | 8 € |
| Gas | 40.0 kWh | 0.05 € / kWh | 2 € |
| Water and wastewater fee | 0.09 $m^3$ | 3.5 € / $m^3$ | 0 € |
| Waste treatment | 7 kg | 4 € / kg | 28 € |

*(continued)*

| | Amount | Cost per unit | Costs |
|---|---|---|---|
| Other services | — | — | 15 € |
| Labor (other) | 1.3 h | 25 € / h | 33 € |
| Depreciation and tax | — | — | 20 € |
| Sum | | | 106 € |
| Total | | | 342 € |
| **Private Household** | | | |
| ***Purchase washing machine*** | 1 | 500 € | 500 € |
| Water | 70.17 m³ | 4 €/m³ | 281 € |
| Electricity | 1117 kWh | 0.18 €/kWh | 201 € |
| Detergents | 183.84 kg | 1.76 €/kg | 324 € |
| End-of-life costs | 1 | 0.00 € | 0 € |
| Sum | | | 1,306 € |
| **Maintenance** | | | |
| ***Maintenance of washing machine*** | 1 | 10 € per annum | 110 € |
| Sum | | | 110 € |
| ***End of life*** | | | |
| Collection | 1 | 8 € | 8 € |
| Disassembly | 1 | 16 € | 16 € |
| Disassembly revenues (reuse) | 1 | –48 € | –48 € |
| Recycling | 1 | 5 € | 5 € |
| Recycling revenues | 1 | –15 € | –15 € |
| Sum | | | –34 € |

*Source:* Real case study (main cost categories, 3rd and 4th level, from Rüdenauer and Grießhammer [2004]; Kunst [2003]) with hypothetical extensions (some cost categories of manufacturer and end-of-life service provider).

## 2.3.4 Cost Estimation

Cost estimation is, quite basically, "the act of approximating the cost of something based on information available at the time" (US Department of Defense 1999). For LCC applications, the "something" may be the product or product components for a certain part of the life cycle or actions and processes in the life cycle such as human labor. Cost estimation implies an assessment of the value or price something has. In comparison to a measurement or calculation of material flows, as is needed for example in an LCA that forms part of an environmental LCC, there are 2 important differences: first, the value will to some degree be volatile; and second, as far as internal costs are concerned, the value will to some degree be publicly available via market prices.

In conventional LCC, a top-down and a bottom-up approach are often used in parallel for cost estimation (e.g., Kerzner 2001). In the top-down approach, costs are derived from an analysis of major components of the product and/or its life cycle. In the bottom-up approach, costs are aggregated from various sources. The variety of cost estimation methods may be classified into informal and formal methods.

Informal methods include expert judgment, analogy, estimation based on relative information, rule-of-thumb methods, the use of engineering standards, and parametric cost estimation.

In parametric cost estimation, the aim is to model a unit (or a "something") in a way that the costs for this unit depend on parameters that can be assessed (more) easily, and with a better estimation quality (Heemstra 1992; US Department of Defense 1999). One example for a parametric cost model would be the effort in person-hours needed for a product development process, based on the type of company, the size of the team, and the "novelty" of the task. These person-hours will then be transformed to cost data by multiplying them with hourly wages.

More sophisticated cost models take into account the nonlinearity of costs. For example, Barry Boehm's famous constructive cost model (COCOMO; Boehm 1981) is, in its basic form, effort = C * $size^M$, where "effort" = person-months needed for a software project, "size" = number of persons in the project group, and C and M are always greater than 1 (for best-practice projects, C = 3.6 and M = 1.2). For environmental LCC, literature on cost estimation is scarce, and costs will often be assessed based on (linear) price–amount relationships. Societal LCC studies may use monetization techniques such as willingness to pay or contingent valuation.

## 2.4 COST BEARERS

Costs involve obligations to pay (or be paid by) legal entities that are involved, including firms, governments, and public bodies. Therefore, the term "cost bearers" refers to those who have to pay the costs that accrue to them. Firms and other organizations may further break down the units that bear costs, for example in divisions, ministries, and associations, as for wastewater management. The duality of cost specification is directly related to who is specified as the cost bearer. A limited number may suffice for total cost specification in the system. The internal cost of these few cost bearers, and all external costs covering their (not overlapping) upstream costs, will be sufficient.

A more encompassing system definition will imply a larger group of activities and, therefore, a larger number of cost bearers. From the point of view of a particular firm, a distinction will be made between downstream proceeds toward the consumer and beyond, and upstream costs in supplying materials and parts (e.g., to the manufacturer). These downstream and upstream costs are related to the life cycle of the product: "upstream" means earlier in the life cycle, whereas "downstream" means later in the life cycle, relative to some reference activity. For instance, upstream from a convenience store are producers, while downstream are consumers and waste-recycling and -processing companies. Eight types of cost bearers may be distinguished, as shown in Table 2.2.

The costs of a producer are essentially the costs of manufacturing a good or service. Costs from producers upstream are counted as long as they are reflected in the price of the purchased goods used as inputs. This may not always be clear in the case of combined production (several products being produced together) when cost allocation rules as applied may differ and may be inappropriate. Related to the producer is the supply chain, which can include all actors from extraction to retail (if the producer is a retailer). For a supply chain, all costs upstream will be taken into

**TABLE 2.2**
**Overview of cost bearers and relevant costs covered**

| Cost bearer | Upstream cost (cost of purchases) | Internal cost (value added) | Downstream cost (*not* subtracting proceeds of sales) |
|---|---|---|---|
| Supply chain | Price | All | None* |
| Producer | Price | All | None* |
| 1st to *n*th owner and/or user | Price | All | Residual value |
| Last owner and/or user | Price | All | Disposal fee, if any |
| Group** | Almost all | All | Almost all |
| Life cycle (all stakeholders) | All | All | All |
| Country's society | All | All | All |
| Global society | All | All | All |

* Only cradle-to-gate costs, unless EoL costs, are part of the company's costs.

** For example, waste collectors and recyclers, excluding consumer costs for separate collection.

account, but downstream costs are taken into account only if EoL costs are part of the company's costs.

Two additional related cost bearers are owners and users. An owner may also be a user, while a user may not be the owner. All upstream costs, reflected in the price of the good or service (either rent or purchase price), will be included. Furthermore, from a full life cycle perspective, downstream costs would have to be included as well, even if not paid by the firms from whose perspective internal costs are being defined.

Groups may be combinations of persons and organizations relevant in a certain situation. One example is the group of users and suppliers of a service, as those involved in car leasing. Groups, as a flexible category, may overlap with any of the other categories. A specific group concerns all actors involved in the life cycle stages of a good or service, from extraction and production to use and disposal; that is the life cycle of the product, where all downstream and upstream costs are analyzed, including cost such as infrastructure overheads and public waste management. This, again, is the full life cycle. It is clear that all partial systems, not covering the full cycle, lead to unclear system boundaries. Unclear definitions of internal and external costs may easily lead to overlapping or missing out costs. Internal and external costs, and the means to categorize and use them, are discussed at length in Chapter 4.

The last 2 groups of cost bearers are a country's society and the global society. The country's society excludes the costs abroad. The view of global society, related to cost bearers, is the most relevant one from a sustainability point of view, since most cost effects (and environmental impacts) do not stop at the border. Case Study Box 3 illustrates the different perspectives discussed herein.

## Case Study Box 3: Perspectives

Starting from the complete life cycle cost result for the idealized washing machine, this case study box illustrates life cycle costs from different perspectives for systems of various persons or groups. In this example, the producer is responsible for the production costs, the maintenance costs for the 1st two years of the use phase (warranty), and the end-of-life costs except for collection (here, disassembly and/or recycling costs minus reuse and/or secondary materials revenues), which result in 320 € over 13 years. The consumer (private household) bears the costs of the use phase, except the maintenance costs for the 1st two years as the warranty usually covers this period. Similarly, the end-of-life costs are shared according to the WEEE (waste electrical and electronic equipment) directive by the producer (disassembly and/or recycling) and public bodies (collection) (European Union 2003a). Next to these conventional costs, monetized externalities could be considered in a societal LCC (e.g., environmental damage costs for emissions) that are borne by the government and society today and in the long-term future.

The environmental LCC for this washing machine considers end-of-life revenues and results in 1216 €, comprising the costs from the perspective of the producer (320 €) and from the perspective of the private household (896 €). Externalities and other costs, like collection costs for old washing machines, are costs covered by the government and society, resulting in an additional 575 € for a societal LCC (1791 €).

Please be aware that prices, hence individual profits or even losses, are not considered in this idealized calculation and have to be added in any real-life study.

| Years | Production | Use | Maintenance | End of life | Externalities |
|---|---|---|---|---|---|
| | A | B | C | D | E |
| 1 | 342 € | — | — | — | 103 € |
| 2 to 12 | — | Water: 25.50 € per annum<br>Electricity: 18.30 € per annum<br>Detergent: 29.50 € per annum | 10 € per annum | — | 42.20 € per annum |
| 13 | — | — | — | Collection: 8 € | 0 €* |
| | | | | Disassembly and/or recycling: 21 € | |

*(continued)*

| Years | Production | Use | Maintenance | End of life | Externalities |
|---|---|---|---|---|---|
| 13, cont'd | | | | Reuse and secondary materials revenues: –63 € | |
| Total, without discounting | 342 € | 806 € | 110 € | –34 € | 567 € |

*Note:* Years: 1 = production, 2 to 12 = 11 years of use, and 13 = end of life.

| Years | Producer | Private households | Government/society |
|---|---|---|---|
| | A+C+D | B+C | D+E |
| 1 | 342 € | — | 103 € |
| 2 to 12 | 20 € | 896 € | 464 € |
| 13 | –42 € | — | 8 € |
| Total | 320 € | 896 € | 575 € |

*Note:* Years: 1 = production, 2 to 12 = 11 years of use, and 13 = end of life.
[*] End-of-life costs and savings related to externalities are assumed to balance each other.
*Source:* Real case study (consumer perspective from Rüdenauer, Grießhammer 2004) with hypothetical extensions (perspectives of manufacturer and for government and society).

## 2.5 UNCERTAINTIES AND INCONSISTENCIES IN COST DATA

Inconsistencies in the costs used in LCC can relate to the definition of the cost collection methods, geographical differences, exchange rates, as well as confidential information, among others.

Finally, should the costing be carried out for publicly available comparisons (as in ISO 14040/44 [2006] for environmental comparative assertions), some internal data are unlikely to be employed and the back-calculation of costs from market prices and value added is an approximation at best, though required.

### 2.5.1 Definitions of Cost Collection Methods

The issue of definitions arises because costs may be defined in different ways. What really is accounted for when a cost is given for a certain good depends on the cost management and accounting system of the reporting party. It further depends on whether the cost is to be used only internally or also for communication outside the organization.

*The accounting system*: There are several diverging accounting systems, for example the generally accepted accounting principles (GAAP) used in the United States, Canada, or the United Kingdom; or the international financial reporting standards (IFRS), used in many parts of the world, including the European Union, Hong Kong, Australia, India, Russia, South Africa, and Singapore. Companies that need to publish their business reports follow one or another accounting system, by legal requirements. Differences between different accounting systems possibly affect the cost given for a certain good or product, the allocation of costs to different cost drivers, and the total of costs allocated.

*The type of cost reported*: It is common practice in many companies to have an internal cost management system that reports, for example for the company's control system, costs independently from legal requirements. The company-internal system is not regulated in any way; there are only common rules of good practice. In the end, each company is free to install its own cost management and cost-reporting system and to use one or the other cost assessment method.

A format for cost data that ensures that companies along a supply chain report the same type of cost, documenting how costs are allocated, promises to solve both the definition and the "cost-type" issue, in case it is applied throughout the whole chain. However, it seems to be difficult to arrive at a uniform format that can accommodate all needs (Rebitzer 2005), and this objective appears rather theoretical.

Using activity-based costing can help to reduce the share of overhead costs and thus the share of costs that need to be allocated; thus, it promises to arrive at more consistent and comparable cost figures.

An employee, even if his responsibility comprise LCC, will rarely be able to calculate all costs in a different way than the company's cost accounting system. A second best approach is, then, to state what types of costs are included in the data given and reported. Nonmonetized costs, such as those derived from surveys indicating willingness to pay, are highly uncertain.

### 2.5.2 Geographical Differences and Exchange Rates

In LCC, which certainly involves global impacts and costs, the exchange rate variations render the final result of the costing time-sensitive. Time and space change the amount of costs. In a different location, identical products may be of completely different value, and costs may need to be paid in a different currency, with floating exchange rates. In a different time, prices and costs may be different. As an example, in December 2001, 0.9 € equaled US$1, while in July 2003, the ratio was 1.15 € to US$1 — about a 30% difference in less than 20 months. In November 2005, 1 liter of petrol cost US$2.21 per gallon in the United States (US Department of Energy 2005), which equals about 0.49 € per liter, and 1.22 € per liter in Germany. A positive discount rate can address future uncertainties (see Section 2.6.1).

Costs in different regions worldwide may be collected for an effective day and transformed into 1 currency. Costs incurred at different times can be stated as such.

### 2.5.3 Confidential Information

The profit earned by selling a product is vital information for any company and is at the same time highly interesting for the company's competitors. Given that the price is often publicly known and that the profit is calculated by price minus costs, one can understand that cost data are often sensitive information. In the not so long history of life cycle costing, several approaches for overcoming the "confidentiality issue" have been proposed.

A common way is stating a "price = cost" equivalency. Since the price is publicly known, it is rather easy to collect. In general, the price is higher than the costs if these exclude profits; in some cases, it may also be lower, as when losses are made. A product's price is also very convenient for estimating the costs of the whole supply chain of the product.

The VDI (Verein Deutscher Ingenieure or Society of German Engineers) has proposed a relative pricing system that allows stating the costs of a product in relation to a basic steel in relation to a basic, nonalloyed steel. Costs for this steel are assessed, published, and updated. Relations of these cost data to other materials such as metals, and to the volume and geometry of the product, are also provided (VDI 2225; Verein Deutscher Ingenieure 1984).

## 2.6 COST AGGREGATION

The last dimension of LCC concerns the manner in which the different costs, revenues, and benefits are aggregated. Though costs are unambiguously summed, unlike environmental impacts, the selection of the appropriate indicator (e.g., net present value) and the decision as to if discounting should be carried out merit consideration. Further, one must also determine if a total cost over the life of the functional unit, or a normalized (e.g., annual) cost, should be employed. The latter is particularly important if 2 alternatives have different lifetimes and/or different operating costs or EoL scenarios. This section will evaluate discounting for each of the 3 types of LCC.

### 2.6.1 Discounting

The reasons for discounting depend very much on the question to be addressed. In conventional LCC, an individual firm may want to know if a profit can be made on a technology choice. It then at least has to deal with the real cost of borrowing. This market rate reflects the reliability of the firm (though the investment could also be financed out of equity, and then the earnings–price ratio for the sector and firm would define the discount rate). Some firms, such as in the information technology (IT), biotech, or pharmaceutical sectors, may have profit rates on investment above 20%, and hence should reckon with this rate. Typically, the discount rate for private investments is between 5% and 20%, to be decided by the private decision maker. For long-term projects in the public sector, such as utilities, the discount rates can be as low as 2%. For societal LCC, the question is how society would evaluate the postponement of costs or benefits. Discounting of the LCC result (note that this is different from discounting cash flows within the calculation procedure; see below) is inconsistent with the steady-state environmental LCC, and, as such, environmental LCC must present its results, for comparison with the long-term effects of LCA, in a time-invariant manner. However, the use of discounted cash flows for money flows occurring at different times within 1 product life cycle (usually for periods no longer than 5 to 15 years, similar to the depreciation period) is commonly applied and does not violate the steady-state assumption.

#### 2.6.1.1 Long-Term Discounting of Costs and Environmental Impacts in Societal LCC

When analyzing the cost of a product system, it is tempting to use one (high) discount rate for economic calculations and another, low one (often 0) for environmental impacts. There are also advocates for a declining discount rate, beginning with an economic one (e.g., 10%) and phasing in, over the economic life, an environmental one (e.g., 0.01%). One should also evaluate, for societal LCC, if discounting should be the same for various environmental impacts. To discuss this issue, which can be elaborated upon in depth (see Howarth 1995; Hellweg et al. 2003), it is useful to look at some typical environmental impacts that often dominate LCA.

Climate change has a number of outcomes around a most likely middle value, with low probability options in terms of runaway effects. Induced climate changes will last for several centuries, while the influences of climate change, including sea level rise, will last longer. The effects on nature in terms of biodiversity loss will last for as long as it takes to develop new species. A time horizon of a million years seems beyond what anybody would reckon as relevant. However, a time horizon of fewer than 1000 years seems reckless from a concerned point of view. Using a discount rate of 0.1% halves the importance of effects every 700 years. Therefore, the 0.01% rate seems an order of magnitude not to surpass to keep the environmentally concerned stakeholders on board.

Toxicity effects have a profile that is quite high at the outset, then decreases. However, in current timeless fate models within life cycle impact assessments, which also disregard natural background concentrations, heavy metals move to the oceans and remain in solution there for well more than a million years. By adding up the exposure times of individuals, weak effects in reality might still become dominant in the LCA. There is concern that calculating costs now for a future population much further away in time than humankind has lasted (let's say 50000 generations) is senseless due to many uncertainties as indicated in the questions and answers above. For example, with inexpensive solar energy and hyperfiltration, the metals in the ocean may be a valuable stock to be mined, with depletion a problem instead of contamination. Using a 0.1% discount rate will fade out the 1 million years' effects to 0. Therefore, toxicity requires a rate of 0.001%. On the other hand, toxicity models in life cycle impact assessment (LCIA) will also have to take time effects such as deposition, which removes toxics from the biosphere, into account in order to appropriately assess long-term environmental impacts.

Within abiotic resource depletion, extracted elements clearly do not remain underground, though they are still part of the mass of the (eco)system. With metals, for instance, there will be 1 part much more highly concentrated than in ores, and 1 part dissipating to very low concentrations. Current depletion scores are very difficult to link to time series. One approach is to couple depletion to the increased energy cost of producing the resource as due to lower concentrations to be mined, as in the environmental priority strategy (EPS) system (Steen 1999a, Steen 1999b). The weakness of such an approach is that it assumes price levels and technologies to be constant for a long time to come — essentially infinitely. The historical fact of the last centuries has been that technologies have developed rapidly, and have led to a systematic decreasing of prices of primary resources (though there are exceptions), apart from short-term fluctuations, even with resources being constantly consumed. Therefore, with abiotic depletion, the basis on which to start the discounting may be weak. Using commercial rates will limit the time horizon of the effects to several decades; using the societal rate of 0.1% will give a time horizon in the order of a few thousand years for this effect (though the appropriate rate may also very much depend on the specific resource that is depleted), which seems reasonable in the case of resource depletion. A longer time frame for resources is not necessary, since a certain level of depletion and thus increase of cost will always lead to adapted technology developments and substitution.

The aforementioned discussion indicates that the determination of an appropriate discount rate for societal LCC is iterative and requires a sensitivity analysis. One must define the predominant impacts and use the discount rate that is the lowest of any acceptable for the above-threshold impacts. Using the examples above, if resource depletion (0.1%), climate change (0.01%), and toxicity (0.001%) were significant impacts in a given societal LCC, one would be obliged to use the 0.001% rate for discounting of the externalities. The reason is that if one were to apply the discount rate for resource depletion (0.1%) to the costs of climate change and toxicity, this would render the midterm effects of these impacts worthless.

The necessity of selecting the lowest discount rate from those available can best be appreciated by examining how installations depreciate. An electrical power plant may have a useful life of 30 years, and therefore the straight-line depreciation could be 3% per annum. Much of the major equipment in the installation has a useful life of 10 years and would be depreciated at 10% per annum, while smaller equipment such as pumps would no longer function after 5 years and be depreciated at 20% per annum. Should a depreciation rate of 20% per annum be applied to the infrastructure itself, then any costs after 5 years (i.e., between years 6 and 30) would essentially be discounted to 0. The analogy of depreciation permits one to appreciate why a sensitivity study applied on the discount rate, given the various impacts, is necessary. Just as is the case for the electrical power plant, if one were to use the resource depletion discount rate (0.1%) for toxicity, all effects after 1000 years would be set to 0. While this seems a long horizon, for toxicity, impacts may occur after tens of thousands of years (e.g., for radioactive substances). Overall, the selection of a discount rate is 1 of the most critical parameters in an LCC, and the discount rate selected must be appropriate for the case at hand. Further, if the LCC is to be publicly disclosed, the discount rate selected will have to be part of the evaluation (and possibly the external review work in analogy to the review requirements of ISO 14040/44 [2006] relating to comparative assertions intended to be disclosed to the public).

The reader should note that the aforementioned reasoning holds for societal LCC only. For conventional LCC, the market, or equity, rates are the starting point. The conclusion may well be that it is normally useful to engage in both types of analysis. The market-based analysis, with high discounting rates, shows the private cost and profitability of options, which may be used for establishing the private costs of certain environmental improvements. The societal analysis indicates how a trade-off between the welfare effects of market effects and nonmarket effects can be made.

### Case Study Box 4: Long-Term Discounting of Results

This case study box illustrates the use of different rates for long-term discounted environmental impacts, based on the idealized washing machine case.

To render a comparison possible between costs and different environmental impacts, the values indicated for the washing machine in Case Study Box 7 have been set at 100% (environmental LCC: 1216 €, toxicity: 0.001 kg benzene equivalent, climate change: 1657 kg $CO_2$ equivalent, and abiotic resource depletion: 830 kg oil equivalent). For purposes of demonstration, no predominant impact has been defined, resulting in the use of the lowest discount rate, but all discount rates deduced in Section 2.6.1 have been applied: the use of a rate of 0.1% for abiotic resource depletion halves the importance of the effect every 700 years, whereas the rate of 0.01% for climate change halves the importance only every 7000 years. The low rate of 0.001% for toxicity reflects the 1-million-year effect of this environmental impact.

In contrast to the long-term discounting of environmental impacts, the costs result of environmental LCC is not discounted as recommended (see Table 1.1).

*Source*: Real case study not available; hypothetical discounted results.

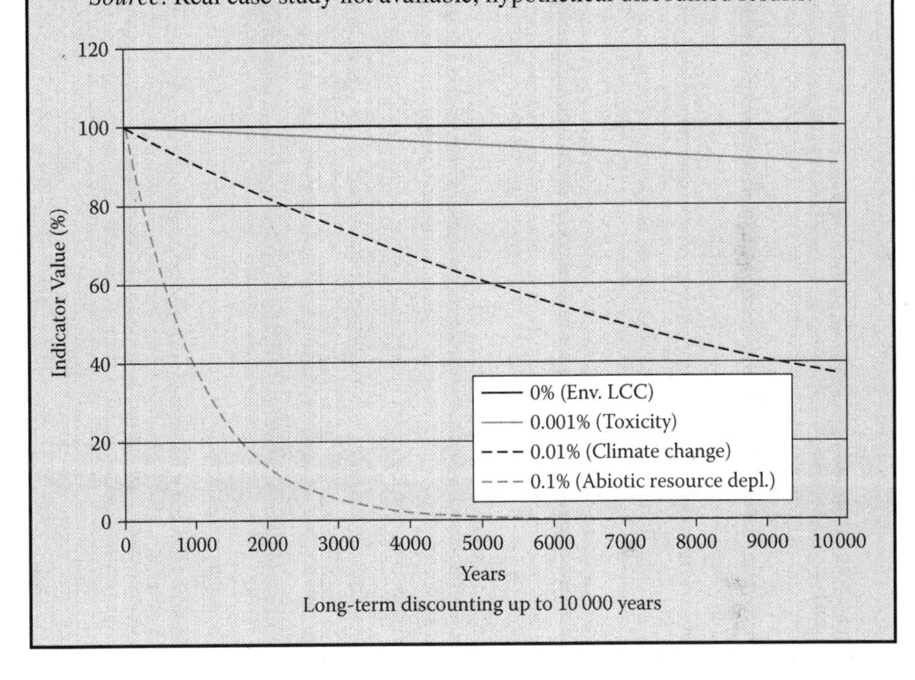


Long-term discounting up to 10 000 years

Most economists advocate that this diversion is not tenable, though practical solutions are lacking. One direction to look at is the (unwanted) steady state that the Japanese economy has been in for approximately a decade, with real discounting rates indeed close to 0. Table 2.3 summarizes the SETAC-Europe working group's recommendations on discounting in LCC.

**TABLE 2.3**

**Summary of recommended discounting of the life cycle costing results**

| LCC types | Environmental LCC | Societal LCC | Conventional LCC |
|---|---|---|---|
| Discounting of results | Inconsistent and not recommended | Recommended | Recommended (though usually not applied) |
| Discounting type | Discounting of the result is not usually possible or easily done as activities leading to inventory results are not normally specified in time, and impacts in the environment are not specified in time.<br>If discounting is carried out, it should be done as described in either conventional or societal LCC, as is the most relevant to the case at hand. Both discounting approaches could be relevant in 1 particular case, implying that 1 case could use 2 discount rates. | Use a social rate of social time preference type for discounting rate of cost and benefits of economic activities (results of LCC). There are several such discount rates, as discussed in Section 2.6, and these go in the direction of the Brundtland Commission's (1987) requirements, with an approximate value below 0.1% per year though likely 1 or 2 orders of magnitude below this, depending on the impacts. It is also possible to distinguish discount rates for different time horizons, going to very low rates for very long horizons. Use this both for economic cost and benefits and for environmental effects and externalities. | Use a market-based discount rate, typically 5 to15%, for cost and revenues of economic activities (results of LCC). Do not apply this to environmental effects and externalities. The lowest rate to be applied would be the market equity rate for a firm in a given sector, corrected for inflation, and the upper range would be the internal rate used by organizations for their intended *return on investment*. This choice is up to the decision maker. The lower limit, clearly, differs geographically.<br>For environmental impacts of LCA type, no time specification is present. This might be developed, similar as for the CBA type of approaches. |
| Remarks | Time specification of inventory activities is at variance with the basic assumptions of LCA. | The assumption is that endpoint modeling for external environmental effects is good enough for this purpose and is capable of being specified in time. This is the case only very partially. The practical difficulties in modeling for discounting are explained with 3 examples in Section 2.6: climate change, resource depletion, and toxic substances. | If applying the CBA type of discounted externalities approach, do not mix the economy-oriented scores and those on external effects, as they are specified on different assumptions. They just mean something different and therefore cannot be added. |

*Note:* This must not be confused with the use of discounted cash flows within the time frame of the product life cycle, which, in any LCC type, depends on the goal and scope and the duration time of a product life cycle.

# 3 Environmental Life Cycle Costing

*Gerald Rebitzer and Shinichiro Nakamura*

**Summary**

The rationale behind environmental LCC is presented, with a specific focus on key issues that one must consider prior to, and during, the assessment. Specific discussions on the appropriate system boundaries, as well as other links to life cycle assessment, are discussed. These methodological issues include the definition of the functional unit and the most appropriate means for data aggregation. The interpretation of the results and the use of portfolio presentations of LCC as a function of the key environmental impact are recommended. Input–output-based LCC is also presented and applied to the cross-cutting washing machine case.

## 3.1 OBJECTIVES OF ENVIRONMENTAL LCC

Environmental LCC is an approach to estimate the economic dimension alone or as part of a sustainability assessment (see Chapter 9).* Therefore, as for the environmental assessment, it is of utmost importance to provide an assessment that can be quantified and thus be used for measuring progress. Without metrics and thresholds, aspects of sustainability cannot be managed and thus improved. It is assumed that the environmental dimension is covered by LCA methods and the social aspects by other approaches, which, however, are in the early stages of development (Klöpffer 2003; Hunkeler and Rebitzer 2005; Weidema 2006).

It should be noted that the environmental LCC methodology is usually meant to be used for validated, though approximate, cost estimations in, for example, product development or marketing analysis. With its comparative and systemic nature, aimed at decision making in the sustainability context, it does not replace traditional detailed cost accounting or cost management practices. It is, rather, a specific, defined, and to-be-standardized tool to estimate decision-relevant differences between alternative products, based on real monetary flows, or to identify improvement potentials within 1 life cycle. One can also observe, in reference to LCA terminology, that the

* Sections 3.1 to 3.3 are largely based on Rebitzer (2005), though in this book the new terminology "environmental LCC" is used instead of "life cycle inventory (LCI)–based LCC," as in Rebitzer (2005), with both nomenclatures being synonymous.

LCC method presented herein aims, primarily, at a consequential approach, and thus resembles LCC planning (see Chapter 1). However, it can also be used for LCC analysis (similar to the attributional LCA approach) if the required scope (e.g., reporting and learning purposes) is met. For a discussion of the attributional and consequential approaches of LCA, which can be transferred to LCC, see Rebitzer et al. (2004) and Ekvall and Weidema (2004).

In general, environmental LCC aims at

- comparing life cycle costs of alternatives;
- detecting direct and indirect (hidden) cost drivers;
- recording the improvements made by a firm in regard to a given product (reporting);
- estimating improvements of planned product changes, including process changes within a life cycle, or product innovations; and
- identifying win–win situations and trade-offs in the life cycle of a product, once it is combined with LCA (and, ultimately, societal assessments once standardized or consensus methods are available for this pillar of sustainability).

## 3.2 SYSTEM BOUNDARIES AND SCOPE

### 3.2.1 Market Structure, Environmental Taxes, and Subsidies

The terms and boundaries for economic systems, as well as for social and natural systems, are not synonymous with those of the product system in LCA. For a common assessment of 2 or 3 of the sustainability pillars, the product system has to have equivalent system boundaries (as stressed by, e.g., Klöpffer 2003; Schmidt 2003). If one examines a perfectly free market, without any environmental taxes or subsidies to account for externalities, environmental LCC could focus only on the economic system provided the following condition is satisfied. Environmental LCC must be applied in conjunction with environmental and/or societal assessments for the same product system with equivalent system boundaries. Under such an (albeit simplified) scenario, all externalities are covered by the other assessments within sustainability assessment. On the other hand, if taxes and subsidies exist and they are comprehensive and fair,* or justifiable based on the collection of a social overhead based on a product's burden, then the economic system can be used as a simplification for the complete social and natural system. Therefore, in the ideal case where all externalities would be completely and perfectly covered by tax and subsidy mechanisms, nationally and supranationally, LCC could provide all the necessary information, rendering systematic environmental and other assessments unnecessary for all but new products.

* A simple, though relevant, example is the cost, to the user, of cigarettes. Clearly, the high taxes contribute to the social and environmental overhead of smoking. However, the price of a box of cigarettes, which typically is 4 € in Europe, is a lucrative tax means that may over- or underestimate the actual externalities. If these taxes are comprehensive and fair from public health and environmental perspectives, then the externalities are built in. If they are unfair, then externalities can be unaccounted for, double counted, or otherwise under- or overestimated.

Clearly, the aforementioned economic assumptions are oversimplified, and, in particular, the latter (complete coverage of externalities by tax and subsidy mechanisms) is highly improbable. If one assumes the tax system is valid for certain products, and not so for others, from socioenvironmental perspectives, then integrating externalities (as suggested, e.g., by White et al. 1996; Shapiro 2001) could, theoretically, provide the complementary information needed to consider the social and environmental consequences of a decision. This would lead to a full aggregation of the 3 pillars of sustainability* in monetary units. Though such an aggregation might be desirable from an ease-of-decision-making point of view, it can be contradictory to the goals of making life cycle approaches transparent, understandable, operational, and readily applicable in routine decision making. This is relevant for firms of all sizes, because a full aggregation would drastically increase the complexity of the analyses and introduce additional value choices and major methodological problems of other disciplines, as, for example, macroeconomic cost–benefit analysis (for a discussion of the associated issues, see Chapter 4).

In conclusion, it seems appropriate to base LCC, as long as it is framed by independent other assessments such as LCA, on the assumption of a primarily unregulated market (see above), even if this includes some double counting for the external effects actually internalized via taxes or subsidies and introduces additional uncertainties. Double counting is, clearly, an issue to minimize, though its avoidance in total is unlikely, and one should be aware of instances where it occurs and ensure it is consistent for all alternatives being compared.

### 3.2.2 Product Life Cycle from Economic and Environmental Perspectives

As explained in Chapter 1, the term "life cycle" has to be seen analogously to the physical life cycle for a functional unit, as in LCA. However, while the latter usually includes the phases of production (from raw materials extraction to manufacturing), use and consumption, and end of life (i.e., "from cradle to grave"), the life cycle in LCC may start even earlier since it also may include the "knowledge" phase (e.g., research and development and acquisition via the supply chain). This is not a fundamental difference to the physical life cycle of LCA since R&D activities may easily be included in LCA as well. It is plausible to assume that for most industrial mass products, resources consumed and substances emitted during the R&D phase usually do not have any significant impact on the environmental performance, owing to the fact that they can be allocated to a high quantity of products. In addition, the absolute material and energy flows originating in R&D are rather small, since this mainly involves thought and modeling and calculation processes as well as laboratory and testing work, though no large production volumes. Therefore, one could argue that R&D is also part of LCA, though usually not included, because its direct impact can be neglected (contrary to the influence of the R&D phase on the environmental performance of the other life cycle phases; see Rebitzer 2005).

Other elements that are usually not included in LCA, such as for instance marketing activities, can also be consistently included in the physical life cycle with the

* Environmental, economic, and social aspects form the 3 pillars (see Chapter 9).

same rationale as the R&D phase. They can be viewed as part of the production phase that is neglected in LCA due to the normally irrelevant influence. However, if for instance a marketing campaign causes relevant environmental impacts, this should also be within the system boundaries of LCA. The same rationale applies for infrastructure and machinery, which is often excluded in LCA because it is seen as negligible, although it is often very relevant in LCC. Also here, it is not a question of inclusion or not, but rather the issue of if the resulting effects on costs or environmental impact are relevant for the assessments. Therefore, given that thresholds will exist for any economic or environmental assessment, LCC and LCA are consistent, though different elements fall below the generally acceptable cutoffs of approximately 5% (Rebitzer 2005).

One can note that additional elements that are of interest from the economic, though not the environmental, perspective can be included without violating the framework condition that the boundaries of LCA and LCC should be equivalent. The same is true for a specific assessment of the environmental and economic implications of a decision. If selected parts of the system are not taken into account in the economic assessment because they are known to be insignificant, they can still be included in the environmental assessment and vice versa. One could also say that the assessment system (environmental or economic) and the addressed scope (what environmental or economic impacts to include) can be different, though the system boundaries referring to the product system have to be equivalent.

The resulting concept of LCC, in a simplified form with 1 product manufacturer (producer) and 1 product user (owner), is illustrated in Figure 3.1. This figure shows the producer and the user as the central actors in the life cycle. These actors are the driving force for why a product exists at all, the consumer being the one who seeks to fulfill a need (demand pull) and the manufacturer being the one who offers a suitable product and who, sometimes, also creates a desire for the product's utility via marketing (supply push). Therefore, these 2 actors are both directly interested in the LCC performance; other additional actors, such as those dealing with end-of-life activities, only have a secondary function, delivering a service that either the manufacturer or the consumer is asking for. In addition, in LCA terminology the functional unit in LCA and LCC is always seen from the view of the consumer, while the manufacturer usually delivers the reference flow (see ISO 14040/44 2006) and

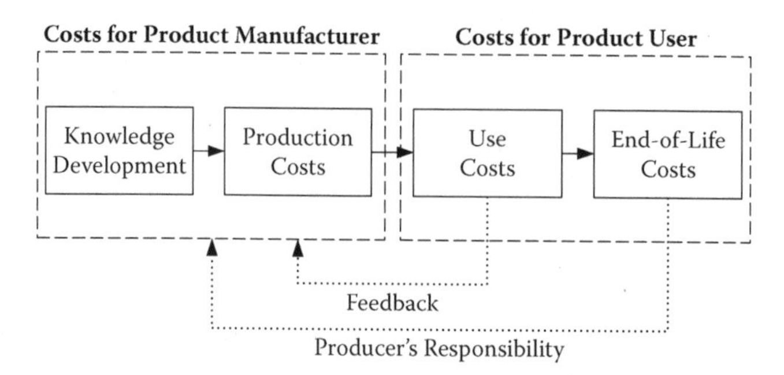


**FIGURE 3.1** Life cycle costing concept. *Note*: Knowledge development can include R&D or, in the case of outsourcing, supply chain coordination. *Source*: Based on Rebitzer (2002).

the EoL actors treat the reference flow after use. This also illustrates that this "LCA terminology" (see ISO 14040/44 2006) can be directly transferred to environmental LCC. If the utility provided by the functional unit is owned by the product user, the LCC approach also resembles the total cost of ownership (TCO) concept.

### 3.2.3 Scope of Environmental LCC

Obviously, the scope of environmental LCC has to differ from that of LCA, since the costs, rather than environmental impacts, are of interest. However, here also connections and overlaps exist. Table 3.1 shows the most relevant costs and how they are connected to elements of LCA. Those costing aspects that can be directly derived from an LCA are written in ***bold italics.*** The life cycle inventory of an LCA provides the quantities of these flows, and the costs can be obtained by multiplying these flows with the respective company costs or market prices (e.g., materials purchasing). Those costing aspects that are written just in *italics* in Table 3.1 can be derived in part or indirectly from the information contained in an LCI. For these aspects, additional information (e.g., the labor requirements for the operation of a certain process) have to be gathered. If this is carried out concurrently with the establishment of the LCI model, minimal additional effort is required, since all processes are studied and analyzed in depth for the LCI. Only the costs associated with research and development (R&D) of the product cannot be derived from the LCA model if the R&D phase is excluded in LCA, which is generally the case (see above). These would then have to be determined separately.

**TABLE 3.1**
**Connection of LCA elements with costs in LCC**

| | **Cost for product manufacturer** | **Cost for product user** |
|---|---|---|
| Production | ***Materials****<br>***Energy***<br>*Machines, plants*<br>*Labor*<br>***Waste management***<br>***Emission controls***<br>***Transports***<br>Marketing activities | Acquisition |
| Use | *Maintenance and repair (warranty)*<br>*Liability*<br>*Infrastructure* | ***Transport***<br>*Storage*<br>***Materials***<br>***Energy***<br>*Maintenance and repair*<br>*Infrastructure* |
| End of life | ***Waste collection, and disassembly/ recycling/disposal if take-back schemes or the like exist*** | ***Waste collection, and disassembly or recycling or disposal*** |

* Categories in *italics* can be directly or indirectly derived from LCA.
*Source:* Modified from Rebitzer (2002).

One can remark that all those processes within the product system that are covered by the LCA are a good basis for deriving the associated costs directly (for material and energy flows) or indirectly (e.g., for labor costs and costs for capital equipment). In addition, only those costs that occur in physical or nonphysical (immaterial) processes, though they are not deemed relevant for the assessment of the environmental impacts, have to be added. This concerns also those costs and impacts, where considered relevant for the goal and scope of the assessment, that are determined via input–output LCA.

The aforementioned links between the product system of LCA with its processes and the corresponding material and energy flows as well as other exchanges (e.g., land use) are the fundamental basis for environmental LCC, which is a life cycle inventory (LCI)–based LCC methodology. For the calculation of the life cycle costs, the same concepts apply whether the product resembles a material good or a service; there are no principal methodological differences.

### 3.2.4 What Environmental LCC Is Not

When discussing the scope of environmental LCC, it is also important to make clear what environmental LCC is not.

Environmental LCC is not a method of financial or managerial accounting (see also Chapter 5). Rather, it is a cost management method within the sustainability framework (see e.g., Seuring 2003) with the goal of estimating costs associated with the existence of a product, just as LCA is not a method of accounting for the absolute environmental impacts of a product, but rather for comparing alternatives. Table 3.2 compares cost management and financial accounting.

Should one seek to better analyze the life cycle costs of a product in detail in order to identify cost drivers and trade-offs for decisions within the life cycle, then existing approaches such as activity-based costing (ABC) can be utilized. For such applications, LCC and ABC complement each other. Even environmental LCC is not intended, nor is it recommended, as a unique tool for sustainability analysis, because it only forms 1 of the 3 pillars of sustainable development.

**TABLE 3.2**
**Comparison of cost management and financial accounting**

| Cost management | Financial accounting |
|---|---|
| Internally focused | Externally focused |
| No mandatory rules | Must follow externally imposed rules |
| Financial and nonfinancial management; subjective information possible | Objective financial information |
| Emphasis on the future | Historical orientation |
| Internal evaluation and decision based on very detailed information | Information about the firm as a whole |
| Broad and multidisciplinary | More self-contained |

*Source:* Hansen and Mowen (1997).

## 3.3 CALCULATING LIFE CYCLE COSTS BASED ON THE PROCESS LCI OF LCA

### 3.3.1 General Procedure

As in LCA, environmental LCC calculations are primarily based on data that are collected per unit process (direct costs). As a unit process (see Glossary) is defined as the single process or subsystem (consisting of several processes) for which data are collected, the level of aggregation can vary highly depending on data availability and the goal and scope of the specific assessment. Indirect costs, such as related overhead costs, can be derived and allocated based on general allocation keys or, in more complex situations, with the help of ABC methods.

Similar to the discussion on the differences between the environmental and economic systems and the boundaries of the product system under study, different levels of aggregation can occur in LCA and environmental LCC, even if both assessments are carried out concurrently for the same product. The desired, or necessary, level of aggregation in LCC depends, aside from the situation of data availability, on the perspective from which the study is carried out (for a discussion of possible perspectives, see Case Study Box 3 in Chapter 2). This means that different unit processes can be used as long as they are compatible to each other (e.g., the material price reflects the complete upstream processes, which consist of many unit processes in the LCI but only 1 subsystem, the cradle-to-gate costs, in LCC). Here, a subsystem denotes a part of the product system model that comprises several unit processes.

The costs for materials and energy and the operation of the processes (e.g., materials and chemicals production, component and product or manufacturing, transport, use, and waste management), as well as additional costs with no equivalents in LCA, must be determined. Subsequently, they are aggregated for the quantity of product (reference flow, derived from the functional unit of the LCA; ISO 14040/44 2006) to be assessed. An example is the aggregation of costs for the treatment of the average amount of municipal wastewater per person and year in a given region (see the case study on wastewater treatment in Chapter 7). For costs or revenues that occur in the mid- to long-term future (e.g., the recycling of an automobile after its useful life 12 years into the future), discounting is relevant. Chapter 2 discusses discounting in the 3 types of LCC.

In addition to defining the reference flows according to the functional unit, which has to be the same as in the underlying LCA model, a cost perspective corresponding to the actor and decision to be supported has to be chosen (see Case Study Box 1, Chapter 1). This is necessary, because the prices are different depending on the perspective due to the value added (including margins) throughout the supply chain. For example, producer prices include the cost of raw materials for the manufacturing of an automobile, whereas consumer prices account for the cost for purchasing a manufactured product such as an automobile (see also Case Study Box 3, Chapter 2 for the washing machine example).

If there are high levels of uncertainty in respect to expected costs, it is advisable to focus on those costs and assumptions that are different in the alternatives studied and to employ sensitivity analyses on a comparative basis. With such procedures, the uncertainty of a comparison of alternatives can be minimized effectively without causing relevant additional efforts for the data compilation process. Only if such an analysis yields high sensitivity of the results to certain data points should specific efforts be undertaken to validate or improve their quality.

### 3.3.2 Specific Methodological Issues: Similarities and Differences between LCA and LCC

#### 3.3.2.1 Definition of Functional Unit and Reference Flows

For environmental LCC, the functional unit has to be the same as in the underlying LCA, because it builds on the same product system providing the same function. While the magnitude of the functional unit might be chosen arbitrarily, it is important to use the same magnitude in LCA and LCC (e.g., packaging for the provision of 1 liter of beverage versus packaging for the provision of the total quantity of beverages consumed by a given population). Therefore, 1 common reference is necessary in order to allow for an appropriate interpretation of the results. In consequence, the reference flows also have to be identical, whether they resemble physical material or energy flows or immaterial services.

#### 3.3.2.2 Definition of Unit Processes, Data Aggregation, and Data Availability

Unit processes and thus the level of data aggregation can, in principle, be regarded as in LCA (i.e., that the data can be collected for the same units). However, in many cases, at least when a detailed assessment of all single technological processes is not necessary, the price for a given process input (e.g., material, component, and service) can serve as a measure for the aggregated upstream costs. In such a case, the detailed costs and added values of the upstream activities need not be known. The implicit cost allocation is based on whatever is used by the firms involved, usually some method of economic allocation like the gross sales value method. This is a fundamental difference to LCA, where data on the complete set of upstream processes are necessary for the calculation of the total environmental impacts, which are not reflected in prices. Therefore, the unit processes do not have to be the same for LCC as for the underlying LCA; aggregates are often sufficient (see also above). On the other hand, if there are cost data available for different unit processes within a product system, they cannot be simply added up as the material and energy flows and/or corresponding impacts in LCA. The value added has to be taken into account, in addition to the costs of purchases of goods and services, for each process. A recommendation is to use market prices for those inputs purchased or outputs for further treatment that are out of the influence or the perspective of interest. Internally, if the aim is to identify cost drivers within 1 organization, costs of inputs and outputs are usually the better choice than market prices. Such choices also reflect the availability of data: costs can often only be obtained from the processes internal to an organization or cost unit, though prices are easily available also from external sources.

A comprehensive example of accounting for cost-related unit processes, divided into meaningful cost categories (though these can vary depending on the study), is given in Case Study Box 2 for the washing machine example (see Chapter 2).

In the context of data availability, it is important to realize that costs and prices can vary greatly over time and from case to case. These depend, for example, on market elasticities, new technological developments, market powers, and transaction costs (for a discussion of the variability of costs and prices and the resulting uncertainties, see Chapter 2). The variance of costs and prices is often much higher than variations in technologies reflected in different LCI data. Therefore, care has to be taken when collecting and using generic cost or price data. Using specific data for the specific object under study, considering the relevant market situations, is highly preferable.

### 3.3.2.3 Allocation in Environmental LCC

Allocation is a heavily debated subject in LCA. In LCC the challenge has a different nature, since coproduct and recycling allocation can be directly done based on market prices. It is obvious to use economic allocation due to the economic nature. However, the allocation of indirect costs such as overheads and the allocation of costs caused by different components within 1 product are important methodological challenges.

The issue of overhead allocation is subject to a complete discipline in economics and can be summarized under methods such as ABC. In this context, environmental LCC can provide improvements since more costs can directly be allocated to the single processes than are usually done in corporate cost management, which is often organized around cost centers without the product perspective in mind. Therefore, environmental LCC can minimize overheads that cannot directly be assigned to single processes and their material or energy flows or other expenses. This can be important, as the survey in Chapter 6 of this book indicates that overhead can often account for more than 50% of the life cycle cost.

The systems view with the focus on processes and products allocates more direct costs by better identifying and eventually transferring indirect costs. An example for this transfer is the cost for the management of production waste, which is often part of the overhead costs of a company, though it can be converted into direct costs by the presented LCC approach. This of course works only if the responsible personnel for waste management have no other obligations in the company that would require, again, an allocation. Allocation thus cannot always be avoided; one has to bear in mind, though, that in environmental LCC, often only those overhead costs that are different from one product to another are of interest — costs that are not product specific can be neglected.

The question of allocating different parts or components (or materials) of a product to costs that can only directly be associated with the complete product has to be solved on a case-by-case basis. An example is the allocation of the weights of different components to the cost of using an automobile. In such cases, where economic allocation cannot be applied, the mechanisms used in LCA should be used. In this example, this would mean allocating the costs of fuel usage responsible for transporting the weight of a component, assuming all other cost-relevant aspects are equal between the alternatives.

### 3.3.3 Use of Discounted Cash Flow

Generally, discounted cash flow is used to calculate money flows occurring at different times of the life cycle of a product. Depending on the assessed product, the goal and scope of the study, and the associated value choices, the discount rate can typically range from 0% to 15% and is occasionally higher. In general, discounting is slightly larger than the local inflation rate. It is recommended to always use a sensitivity analysis (i.e., using different discount rates) in order to evaluate the influence of this methodological choice. If the choice of the discount rate influences the overall ranking of alternatives, this has to be critically discussed, and the associated uncertainties should be mentioned in the interpretation.

### 3.3.4 Data Compilation and Aggregation

There is no generic data format for environmental LCC, and it is uncertain if the data requirements for LCC, in general, can and will be standardized in detail. Data requirements are strongly dependent on the goal and scope of the study, and cost differences are the main concern rather than absolute figures. This also implies that different studies of the same object, with various goals and scopes, are usually not directly comparable to each other, as is the case for LCA studies. In addition, cost information is much more variable over time than life cycle inventory data; therefore, static databases are often not very useful for LCC, while the contrary is the case for flow data of LCI unit processes (for arguments related to the latter issue, see Frischknecht, Rebitzer 2004). However, in cases where specific data are lacking or where only a coarse generic LCC analysis is the goal, prices from databases such as those from Granta Design (2004) can be employed. These data sources provide default price ranges for material as well as manufacturing process costs, or relative cost catalogues (see, e.g., VDI 2225; Verein Deutscher Ingenieure 1984). Furthermore, the field of cost estimation is quite developed and could be used to provide supplemental data to environmental LCC, as is also alluded to in Chapter 5. If environmental LCC is applied regularly within an organization, it is advisable to build and maintain an internal database for the most relevant cost categories of the processes, materials, and energy carriers under study. For the latter case, an internal data format should be established, which should also address issues of currency conversions, fluctuations over time (ranges of prices), and geographical price differences. Such databases could be, and often are for multinationals, quite modular in nature.

The general approach for calculating and aggregating life cycle costs is described in Chapter 2. The specific procedure for information gathering and for identifying and quantifying the relevant cost data per unit process or subsystem of the product system model, and the aggregation to life cycle costs for the production, use, and end-of-life phase in environmental LCC, can be summarized as follows:

*Step 1*) Identification of the subsystems or unit processes that could result in different costs or revenues (in the following steps, only the term "costs" is used, denoting both costs and revenues)

*Step 2*) Assignment of costs or prices to the respective product flows of the unit processes or subsystems identified in step 1, with the process output as a reference unit (e.g., 1 kg intermediate product)

**Case Study Box 5: Calculation with Discounted Cash Flow**

This case study box illustrates the use of different rates for discounted cash flow in **environmental LCC**, using the washing machine case described in detail in Section 7.5.

The energy and water consumption of washing machines decreased to a quite large extent during recent years. In the case study, the costs of the further use of existing washing machines in stock are compared to those of the purchase and use of a new washing machine, which has potentially lower costs during the use phase. To this end, the time required to save the additional costs caused by the acquisition of a new washing machine was analyzed. In addition to the washing process, the drying of clothes was also included, as the energy consumption of the drying process is influenced by the spin speed of washing machines, which also increased during recent years.

Five alternatives were compared: the further use of washing machines of different ages (bought in 1985, 1990, 1995, and 2000), and the acquisition and use of a new one in 2004. For each alternative, the life cycle costs are calculated on an annual basis (per year). These annual values are then cumulated to give the total costs after 1, 2, 3, and, ultimately, up to 10 years of use. Thus it can be determined after what time period the purchase price is compensated by the lower costs during the use phase.

The figures below show the life cycle costs with a rate of 0% for the discounted cash flow (i.e., without discounting) and with a discount rate of 5% to give the net present value (NPV).

*Source*: Real case study (Rüdenauer, Gensch 2005a); no hypothetical extension necessary.

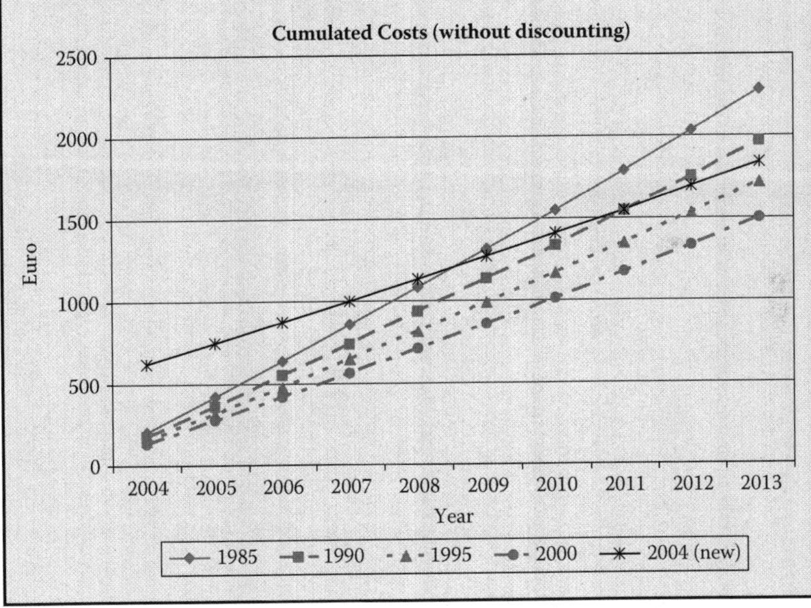

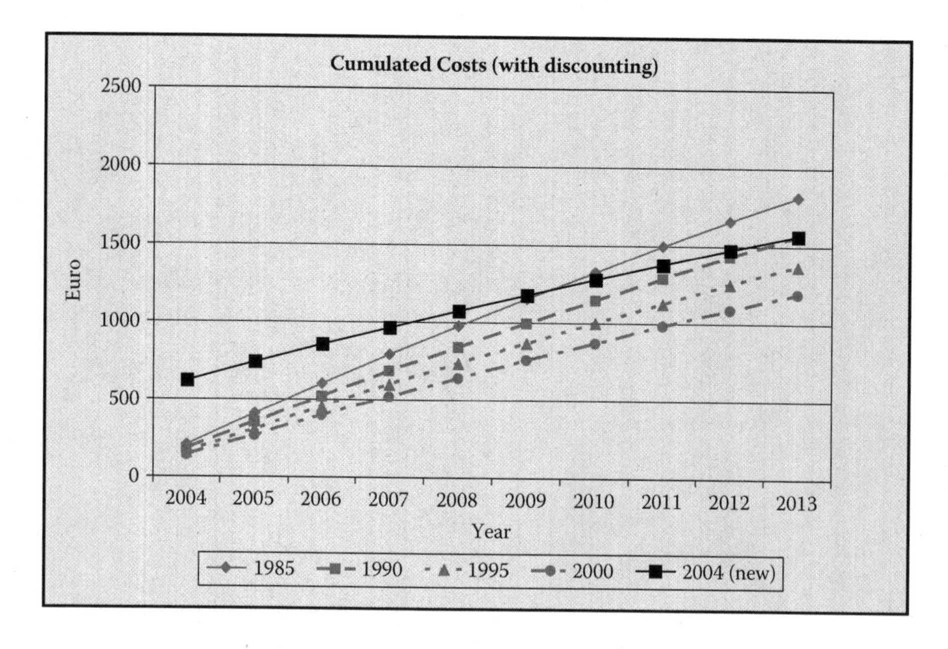


*Step 3*) Identification of additional cost or price effects of the unit processes or subsystems identified in step 1 that differ between the studied alternatives (other operating costs of the process take into account investments, tooling, labor, etc.)

*Step 4*) Assignment of costs or prices to the additional process operating costs identified in Step 3, with the process output as reference unit

*Step 5*) Calculation of the costs per unit process or subsystem by multiplying the costs per reference unit from steps 2 and 4 with the absolute quantities of the process outputs for providing the reference flow(s) of the complete product system

*Step 6*) Aggregation of the costs and prices (from the same perspective, both are outflows) of all unit processes or subsystems (from step 5) over the complete life cycle

Theoretically, the data needed in steps 2 and 4 could be retrieved to a large extent from internal enterprise resource planning software (ERP) systems. These are obtained by coupling these systems with LCI modeling tools (such as, e.g., SimaPro [PRé Consultants 2005] or GaBi [IKP/PE 2004; Ganzheitlichen Bilanzierung 1996]), as suggested for environmental flow data already several years ago (see, e.g., Krcmar 1999; Möller 2000) and since implemented by some corporations (see, e.g., Gabriel et al. 2003). In the future, if LCI-based LCC methods such as presented here become more standard applications, such a coupling of the IT systems should be targeted.

Case Study Box 6 gives some practical guidance on how to carry out steps 1 to 6 to calculate the life cycle costs of a specific product, for example the idealized washing machine.

### Case Study Box 6: Calculation of Life Cycle Costs

This case study box demonstrates how the environmental life cycle costs for the idealized washing machine have been calculated, using the environmental LCC methodology presented in this book.

In step 1, the unit processes of the 3 alternative washing machines resulting in different costs have been identified. In step 2, the costs have been assigned to the respective product flows of each unit process, with the process output as a reference unit. Moreover, in steps 3 and 4 the additional costs of the studied alternatives have been identified and assigned to the subsystem concerned (e.g., investment costs to optimize a washing machine assembly line or a more efficient motor to improve energy efficiency — more copper, less steel).

To calculate the costs related to the life cycle of the washing machine, the following equation has been used for steps 5 and 6 (calculation of the costs per unit process or subsystem = life cycle phase, aggregation of the costs):

$$LCC = \sum_{\text{life cycle phase } 1}^{\text{life cycle phase } n} \sum_{\text{process } 1}^{\text{process } i} \left( \mu_i \times \sum_{\text{cost el. } 1}^{\text{cost el. } p} \sum_{\text{flow } 1}^{\text{flow } q} amount_q \times costs_p \right)$$

where

$i$ = process-specific variable
$p$ = cost category–specific variable
$q$ = process flow–specific variable (can be input or output)
$\mu$ = process scaling factor related to the product system
$n$ = life cycle phase–specific variable

The equation consists of different parts. First, the costs per unit process have been calculated by multiplying the costs per reference unit with the absolute amount of the process outputs for providing the reference flows of the complete product system (e.g., price per kWh × electric energy demand [kWh]). The process scaling factor ($\mu$) indicates which amount of the different processes is needed for the considered washing machine. Then, the costs of all unit processes are aggregated over all life cycle phases (preproduction, production, use, and end of life) to present the costs per subsystem. In the last step, the life cycle cost is calculated by adding the costs of different life cycle phases over the complete life cycle time.

Discounting, where relevant, should then also be integrated into the calculation (see Case Study Box 5, not detailed in the equation above).

This stepwise calculation allows one to provide costs of certain life cycle phases or processes, costs over a certain period of time, or costs for certain cost categories.

Please note that other case study boxes used throughout this book illustrate selected environmental LCC issues using the idealized washing machine and present specific results: for example, Case Study Box 1 on goal and scope definition, Box 2 on cost categories, Box 3 on perspectives, Box 5 on calculations with discounted cash flow, Box 7 on the comparison of 3 types of LCC, and Box 10 on the presentation of environmental LCC results.

### 3.3.5 Interpretation of Environmental LCC Results

In LCA, the interpretation phase is defined as a "systematic procedure to identify, qualify, check and evaluate information from the results of the LCI and/or LCIA of a product system, and to present them in order to meet the requirements of the application as described in the goal and scope of the study" (ISO 14040/44 2006). This definition can be directly transferred to environmental LCC by replacing "of the LCI and/or LCIA" with "of the LCC analysis" (Rebitzer 2005).

The interpretation phase is very specific to a study, involving checks of completeness, consistency, and sensitivity (ISO 14040/44 2006) in order to arrive at findings or recommendations relative to the goal(s). Methods of uncertainty analyses, apart from sensitivity analysis, might also be 1 element of interpretation.

As in LCA, the aim of the interpretation in environmental LCC is to evaluate the results obtained in the LCC, taking into account all previous steps. Uncertainty and sensitivity analysis should focus on those data that might contain the highest uncertainties due to the involvement of coarse assumptions, expected variations (e.g., of elastic market prices or owing to the time dependency of the data for a life cycle that spans several years), or value choices. The latter are always a factor when the discounting of future costs and revenues is applied. If necessary and desired, more sophisticated techniques for assessing uncertainty of cost and revenue input data can also be applied, as demonstrated by Norris and Laurin (2004), who use Monte Carlo analysis for calculating cost originating from risks and liabilities. United Technologies has developed a sensitivity analysis for environmental LCC based on the analytical hierarchy process (Margni et al. 2006).

When interpreting results of LCC, care has to be taken not to underestimate uncertainties, specifically when comparing them to potential variations in LCA. Even though LCC only works with 1 unit (a monetary unit such as dollars, euros, or yen), uncertainties of some costing data might be higher than for technological inventory or impact assessment data (see the discussions in Chapter 2).

In order to identify environmental-economic win–win situations or trade-offs, the final results of an LCC study should be analyzed together with the results of the parallel LCA study. One possibility is to plot selected LCA results (e.g., 1 representative or the [by the LCA interpretation identified] most important impact category) versus the LCC results (portfolio representation, as is presented in the executive summary and detailed in Case Study Box 10 in Chapter 5). If the LCA results show significant trade-offs between impact categories, then it is also possible to create several portfolios. The use of 1-score (i.e., weighted) LCA results is not recommended due to the resulting loss of transparency, acceptance problems, and the requirements of ISO 14040/44 (2006), which directly in regard to weighting only focus on comparative assertions intended to be disclosed to the public, though they are often also followed for other applications.

It is important to note that the aforementioned portfolio presentation only shows relative differences between the alternative products studied in the combined LCA and LCC since both assessments have a comparative nature. This is in contrast to portfolios with similar appearance, which claim to include the economic and environmental impacts of the average good or service. The latter type represents averages relating to the market shares of all goods or services for a given functional unit, placing the average product at the center of the portfolio, as proposed by Saling et al. (2002). Therefore, the resulting portfolio herein is termed "relative life cycle portfolio" (Rebitzer 2005) so that it is not confused with the concept of Saling et al. (2002). In the future, such relative life cycle portfolios should be extended to also include the 3rd dimension of sustainability, social aspects, from a life cycle perspective. The combined results of LCA and LCC can also be used for further analyses in the context of LCM and sustainability. For instance, the normalization to comparable baselines and the subsequent calculation of ratios or metrics can add additional insights to the questions of eco-efficiency and sustainability. Examples of such metrics, where both LCA results and results from environmental LCC can be employed, are the return on environment (ROE; Hunkeler 1999; Hunkeler and Biswas 2000; Hunkeler 2001) and the econo-environmental return (EER; Bage and Samson 2003). ROE calculation applications from the automotive and aerospace sectors involving the environmental LCC method can be found in Rebitzer and Hunkeler (2001). However, care has to be taken for the relative life cycle portfolios in regard to normalization. The use of data and results based on validated LCC and LCA can be invalidated if an inappropriate denominator is used for normalization.

## 3.4 ENVIRONMENTAL LCC IN RELATION TO CONVENTIONAL AND SOCIETAL LCC

One can summarize that environmental LCC resembles the parallel to life cycle assessment, thus being in most cases the most appropriate method for establishing the 2nd pillar of sustainability for product assessments (for a discussion on the role of these pillars, see Chapter 9). In comparison to conventional LCC, environmental LCC includes also anticipated costs and all life cycle steps and is always linked to an environmental life cycle assessment, being based on equivalent system boundaries and a product system model. However, contrary to societal LCC, it does not include externalities that are not borne by any of the actors in the life cycle during the relevant time period (time of the life cycle and period of time during which the product is produced, used, and disposed of). Case Study Box 7 outlines the different calculation results for the case of a washing machine, comparing all 3 LCC types. The results are based on a real conventional case study (Rüdenauer and Grießhammer 2004) with hypothetical additional features of environmental and societal LCC.

## Case Study Box 7: Different Types of LCC

This case study box summarizes the differences in the results obtained, and conclusions derived, from conventional, environmental, and societal LCC. One can observe that the life cycle costs increase as one moves from conventional LCC (1172 €/unit), to environmental LCC (1216 €/unit), to societal LCC (1791 €/unit), reflecting the expanded system boundaries and the inclusion of additional "externalities." This is particularly noticeable in the R&D, preproduction, and production stages, though to a lesser extent also in the use and disposal stages.

The impact assessment results are presented in terms of the 5 main impact categories, which constitute more than 80% of the total burden, are shown for environmental LCC, and are absent from conventional LCC and societal LCC (for the latter, these impacts are part of the monetary values). The dominant impact is clearly the global warming potential with the use phase, which therefore is of particular concern for improvement.

*Source*: Real case study (conventional LCC from Rüdenauer and Grießhammer [2004]; and LCIA from Kunst [2003]) with hypothetical extensions (environmental LCC, societal LCC, and parts of conventional LCC).

| Conventional LCC | | | | Environmental LCC | | | | Societal LCC | | | |
|---|---|---|---|---|---|---|---|---|---|---|---|
| Life cycle stage | Cost (€ per unit) | Principal impact categories | Impact (per unit) | Life cycle stage | Cost (€ per unit) | Principal impact categories | Impact (per unit) | Life cycle stage | Cost (€ per unit) | Principal impact categories | Impact (per unit) |
| R&D | 314 | No complementary LCA required None estimated | | R&D | 20 | Global warming | 1657 kg $CO_2$ equivalent | R&D | 445 | No complementary LCA to avoid double counting (monetary societal assessment and life cycle impact assessment) | |
| Preproduction | — | | | Preproduction | 216 | Acidification | 8 kg $SO_2$ equivalent | Preproduction | — | | |
| Production | — | | | Production | 106 | Eutrophication | 2 kg nitrogen | Production | — | | |
| Use | 858 | | | Use | 916 | Human toxicity | 0.001 kg benzene | Use | 1380 | | |
| End of life | — | | | End of life (with revenues) | –42 | Resource depletion | 830 kg oil | End of life (with revenues) | –34 (collection, externalities costs, and savings balanced) | | |

## 3.5 CALCULATING LIFE CYCLE COSTS BASED ON HYBRID LCA

The integrated use of input–output analysis (I–O) with process data, which is known as "hybrid-LCA" or "I/O-LCA," has evolved into an important tool (Udo de Haes et al. 2004; Suh and Huppes 2005). The I–O methodology that is usually employed in a hybrid LCA is the celebrated Leontief quantity model. The quantity model can compute the sectoral level of output and the associated environmental load that are invoked by a given level of final products. Another equally important I/O methodology is the Leontief price model. The price model can compute the price of sectoral output for a given level of value-added ratios or the costs for primary factors (capital and labor) per unit of output (Miller and Blair 1985). Further, they make use of the same matrix of input coefficients and share the same body of physical and technical information about production processes that constitute the economy. In fact, mathematically speaking, both the (quantity and price) models are dual to each other (Dorfman et al. 1958). Although the physical input–output relationships make the quantity model so useful in LCA, the price model is concerned with the aspect of cost and price, which is the subject of LCC. This is in parallel to the fact that the economic counterpart of LCA is the environmental LCC (Klöpffer 2003). This subsection is concerned with the calculation of the environmental LCC based on the I/O price model.

### 3.5.1 Input–Output Methodology

#### 3.5.1.1 Costs and Prices in Input–Output Analysis

The I–O table is a part of the system of national accounts and is subject to several rules that are necessary to keep consistency within the system. Of these rules, the one that is most important for cost analysis is the following:

the value of output = the value of input, or

the revenue from current production = the expenditure for current input.

For a simple case of the economy that consists of 3 industry sectors, this can be given by

$$p_j x_j = \sum_{i=1}^{3} p_i a_{ij} x_j + V_i, j = 1,\cdot,3, \tag{3.1}$$

where $x_j$ refers to the amount of output $j$, $p_j$ to its price, $a_{ij}$ to the amount of output $i$ that is used to produce a unit of output $j$, and $V_j$ to the gross value added that consists of the cost of primary factors such as capital and labor compensations that occur in sector $j$. This implies that there is nothing like pure or excess profit that is not allocated to any input: all the revenues from current production are completely allocated to each of the input items that have contributed to the production. In particular, corporate profit shows up as a component of capital compensation. Another important implication is that the unit production cost of a product is equal to its selling price, which becomes apparent when both sides are divided by $x_j$:

$$p_j = \sum_{i=1}^{3} p_i a_{ij} + v_j, j = 1,\cdot,3, \tag{3.2}$$

where $v_j \equiv V_j / x_j$ is the value-added ratio. Because $p_i$ occurs on both sides of Equation (3.2), it can be solved for them as follows:

$$p = v(I - A)^{-1}, \tag{3.3}$$

where $p$ refers to the transpose of $(p_1, p_2, p_3)$, $v$ to the transpose of $(v_1, v_2, v_3)$, $A$ to the technology matrix, the $i$th row–$j$th column element of which is $a_{ij}$, and $I$ to a unit matrix of order 3. This is the unit cost or price counterpart of I–O, the price-I–O model (Miller and Blair 1985).

#### 3.5.1.2 Introducing the Use Cost

The costs given by Equation (3.3) refer to the cost in the production phase only (with a possible inclusion of the cost for R&D in annualized form and the cost for marketing). Introducing the costs of the use phase into Equation (3.3) is straightforward. Suppose that the output of sector 1 (henceforth, simply called "product") is a durable product, the use of which over $T$ years requires the input of output 2 by the amount of $b_{21}$ per year. The costs in the use phase can then be introduced into Equation (3.3) by replacing $a_{21}$ in $A$ by $a_{21}^* = a_{21} + b_{21} * T$. The solution $p_1$ then gives the cost of production and use per unit of the product under full consideration of the interdependence among the 3 sectors. Generalization to the case where $b_{21}$ does not remain constant over time can be facilitated by rendering $b_{21}$ time dependent.

#### 3.5.1.3 Introducing the End-of-Life Cost

Introduction of the EoL costs (or revenues) into the I–O framework requires additional consideration of waste and waste treatment activities, which are (explicitly) not present in Equation (3.2). First, the simplest case is considered where the EoL product (the output of sector 1 that is discarded after $T$ years of use) is the only waste, and landfilling is the only waste treatment process that is available. Suppose that sector 3 refers to the landfilling process, with its output $x_3$ giving the amount of waste landfilled, and $p_3$ giving the price of landfilling per unit of waste. This implies that the EoL cost of the product is given by $p_3 a_{31}$, with $a_{31} = 1$, that is, $p_3$. The solution $p_1$ of equation (3.3) with the $a_{21}$ replaced by $a_{21}^*$ and $a_{31} = 1$ then gives the life cycle cost of the product.

One next considers a more general case where the EoL product is no longer directly landfilled, but is subjected to an intermediate treatment process (e.g., disassembly) that separates its feedstock into recyclables and residues. Assume that some portions of recyclables can be recycled in sector 2, while the portions not recycled and residues are landfilled. The disassembling process takes place in sector 4, with $x_4$ denoting the amount of waste processed by this process, and $p_4$ the price of processing a unit of waste. In the simplest case considered above, there were only 1 type of waste and 1 type of waste treatment process. In the present case of multiple wastes (EoL product, recyclables, and residues) and multiple treatment processes, this simple 1-to-1 correspondence between waste and treatment no longer holds. The absence of this 1-to-1 correspondence is a typical situation in waste management, which cannot be dealt with by the conventional I–O. It becomes necessary to resort to the waste I–O (WIO; Nakamura and Kondo 2002).

Table 3.3 provides an extended form of the technology matrix $A$ in the form of WIO. The $3 \times 4$ matrix in the lower half of Table 3.3, with the rows referring to the 3 types of waste, gives the flow of waste per unit of activity, $g_{ij}$, in each of the 4 sectors. $g_{24}$ and $g_{34}$ refer to the amount (say, weight) of waste materials and residues that are obtained from a unit of EoL product per unit of operation of sector 4, and $g_{22}$ to the amount of waste materials that is used (recycled) per unit of output in sector 2.

The presence of recycling implies that it is necessary to consider the price of waste materials and recyclables in cost calculations. The sale of recovered waste materials (for instance, metal scraps from automobile shredding) at a positive price can reduce the EoL cost of a product, while the acceptance of them as input at a negative price (e.g., the burning of waste plastics in a cement kiln) can reduce the production cost of its user. Furthermore, it is also necessary to take into account the cost for managing the wastes that are generated in the production process (Table 3.3 deals with a simple case where there is no waste other than the EoL product). A distinguishing feature of the WIO price model (Nakamura and Kondo 2005) is its full consideration of these effects of recycling and waste management in cost calculations within the hybrid framework.

#### 3.5.1.4 Internalizing External Costs

The internalization of externalities in the WIO framework is straightforward. As an example, consider the case where a carbon tax of $t_C$ per unit of carbon on fuel

**TABLE 3.3**
**Extended I–O coefficients matrix with waste and waste treatment in the form of WIO**

| Input, waste, and output | Sector 1 | Sector 2 | Sector 3 landfill | Sector 4 disassembling |
|---|---|---|---|---|
| Sector 1 | 0 | $a_{12}$ | $a_{13}$ | $a_{14}$ |
| Sector 2 | $a^*_{21}$ | 0 | $a_{23}$ | $a_{24}$ |
| EoL product | 1 | 0 | 0 | 0 |
| Waste materials | 0 | $-g_{22}$ | 0 | $g_{24}$ |
| Residues | 0 | 0 | 0 | $g_{34}$ |

consumption is introduced (in the decision-relevant future according to the goal and scope of the study). Write $e_{Cj}$ for the emission of carbon dioxide (of fuel origins) in carbon weight per unit of producing output $j$. Augmenting the ratio of value-added $v_j$ in Equation (3.3) with $e_{Cj}t_C$ then gives the effects of the tax: the emission of carbon occurs as an additional input of "primary factors of production" such as labor.

### 3.5.2 Numerical Example of I–O-Based LCC for the Washing Machine

For illustrative purposes, the methodology of Section 3.5 has been applied to a simplified case study of a washing machine. It deals with a mere illustrative numerical example and should not be regarded as a comprehensive LCC and LCA study. The functional unit is a washing machine that is used for 9 years and then subjected to an EoL process that is consistent with the Japanese law on the recycling of appliances (Kondo and Nakamura 2004).

#### 3.5.2.1 I–O Data for the Washing Machine Case

Foreground data on the use phase of a washing machine in Table 3.4 are taken from Matsuno et al. (1996). The Japanese WIO table for 2000 provides the basic data, an updated version of the one for 1995 (Nakamura 2003) based on the national I–O table for 2000. The WIO table consists of 396 industry sectors, 3 basic treatment methods (shredding, incineration, and landfilling), and 61 waste types that cover municipal solid waste, commercial waste, and industrial waste. The consideration of the generation and recycling of diverse types of process waste from approximately 400 sectors makes this case more complicated than the preceding case outlined above, requiring the use of a generalized version of the above methodology (see Nakamura and Kondo [2005] for details).

#### 3.5.2.2 I–O Results for the Washing Machine Case

The results, presented in Case Study Box 8, indicate the dominant importance of the use stage in both LCC and LCA (see Figure 3.2). The use stage contributes to more than 60% of the total cost and more than 70% of $CO_2$ and other emissions (this result is consistent with Matsuno et al. 1996). Decomposition of the use stage by each of the inputs indicates that water use and related sewage treatment are the

**TABLE 3.4**
**Data for the washing machine example (1 euro ≈ 142 yen)**

| | Price (€) | Unit weight | Use phase* |
|---|---|---|---|
| Washing machine | 704 | 29kg | 1 |
| Detergent | 2.11 | Kg | 160 |
| Electricity | 0.16 | KWh | 600 |
| Water | 0.73 | $m^3$ | 691 |
| Sewage | 0.54 | $m^3$ | 691 |

* The figures refer to the amount consumed over 9 years.

### Case Study Box 8: LCC via Input–Output Analysis

This case box summarizes the results of the LCC for the washing machine case, carried out using input–output analysis (based on Japanese data). It is evident that costs in the use stage are highest, followed by those in the production stage, while the costs of the EoL stage are almost negligible. Of the 807 kg of $CO_2$ equivalents, 579 kg is derived from the use stage. The same proportions are observed for $NO_x$, suspended particulate matter (SPM), and $SO_x$, with around 70% of the impacts coming from the use stage.

If one compares the results of the I–O analysis (Japanese I–O tables) with the environmental LCC using German data (Case Study Box 7, based on Case Study Box 2) one observes that the costs for production and the EoL scenario are estimated to be much lower using I–O, respectively by a factor of 3× and 100×. The costs during the use phase are somewhat higher for the Japanese case, overall leading to a much higher relative share of the use phase in Japan. This can be explained by the fact that washing is carried out in Japan with a much larger amount of water and hence sewage (wastewater) produced. Water use is about 10 times higher than in Germany (66 $m^3$ water in Germany compared to 691 $m^3$ in Japan). Nevertheless, it should be noted that the water used is cold water as the Japanese washing machines are not equipped with water-heating capacity (often they are operated outside the home) and hence consume less electricity than the German counterparts in the use phase, though these savings cannot outweigh the additional costs for water. The EoL cost corresponds to the current level of recycling fee that is charged when a washing machine is disposed of by a consumer. See Section 3.5.2.2 for details of the data used.

However, it should also be mentioned that some authors have criticized I–O-based LCA for its lack of correspondence to LCI-based LCA. This simple washing machine case illustrates that the differences in I–O and inventory-based LCC exist and may not only be explained by differences in the European and Japanese washing machine technologies and habits as explained above and summarized in Table A.1 in the Appendix to Case Study Boxes. They represent, respectively, macro- and micromeasures of the same phenomena, though the differences, which are important, imply that any user has to be very cautious as to which method is selected.

*Source*: Real case study (Matsuno et al. 1996) with hypothetical extensions.

**Contributions of the various impacts and costs throughout the life cycle stages (1 euro ≈ 142 yen)**

| Life cycle stage | Cost (€ per unit) | Principal impact categories | Impact (units) |
|---|---|---|---|
| R&D | — | Global warming | 807 kg $CO_2$ equivalent |
| Preproduction | — | Acidification | 0.4 kg $SO_2$ equivalent |
| Production | 704 | Eutrophication | 0.6 kg nitrogen |
| Use | 1306 | Suspended particulate matter (SPM) | 0.1 kg of SPM |
| End of life | 17 | — | — |
| Totals | 2027 | — | — |

main components of the cost, while electricity is the main component of $CO_2$ emission, followed by water use and sewage treatment (Figure 3.3). It follows, for this example, that saving water at the use stage is the most effective way for reducing the cost, while saving electricity is most effective for reducing $CO_2$ emission of a washing machine. Table 3.5 provides the background data for the washing machines considered.

Care has to be taken when selecting the method, which should be based on the goal and scope of the study. Process-based LCA and environmental LCC (as described in Sections 3.1 to 3.4) are clearly much more appropriate when doing specific comparisons for specific product models (e.g., in design for environment). On the other hand, if the major goal is to get first an overview about the relevant impacts for a generic product (family) in the sense of a screening application, input–output-based methods can be helpful to provide a 1st analysis — as long as the relevant sectors are well represented and differentiated in the available input–output tables.

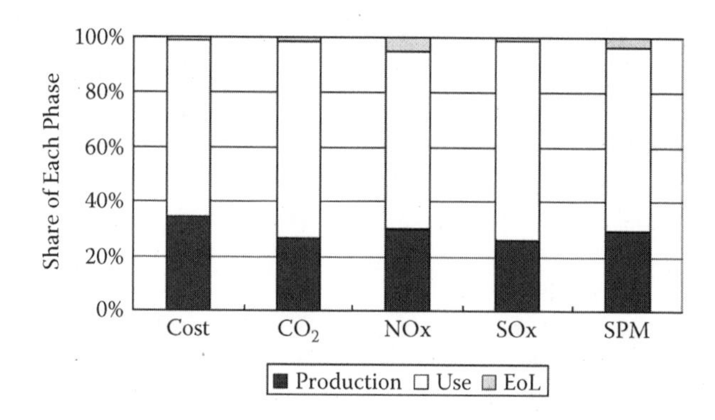


**FIGURE 3.2** Cost and emissions at each of the 3 life cycle phases. *Note*: SPM refers to suspended particulate matter.

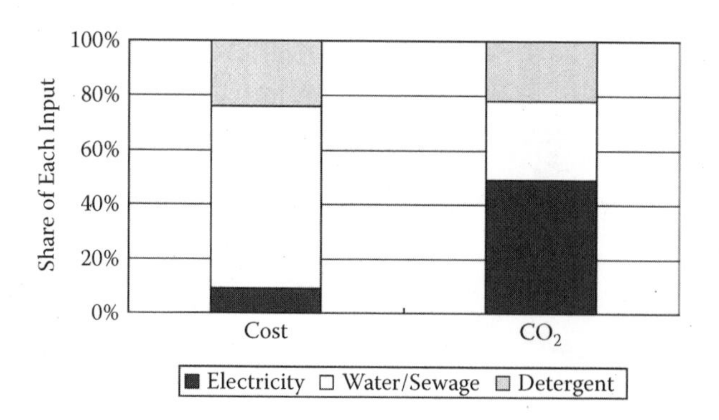


**FIGURE 3.3** Components of the cost and $CO_2$ at the use phase.

**TABLE 3.5**
**Background data on Japanese washing machine (1 euro ≈ 142 yen)**

**Use patterns (Matsuno et al. 1996)**

| | Unit | Amount |
|---|---|---|
| Washing cycles | Per day | 1.4 |
| Washing temperature | | Cold water (no heating) |
| Washer capacity | Liter | 54 |
| Lifetime | Year | 9 |

**Production (Matsuno et al. 1996)**

| | Unit | Amount | Cost (€ per unit) | Costs (€) |
|---|---|---|---|---|
| Metal | kg | 17.24 | | |
| Plastics | kg | 11.82 | | |
| Total | kg | 29.06 | 704 | 704 |

**Inputs in the use phase (Matsuno et al. 1996)**

| | Unit | Amount | Cost (€ per unit) | Costs (€) |
|---|---|---|---|---|
| Water | $M^3$ | 691 | 0.73 | 501 |
| Sewage | $M^3$ | 691 | 0.54 | 370 |
| Detergent | kg | 160 | 2.11 | 338 |
| Electricity | kWh | 600 | 0.16 | 97 |
| Total | | | | 1306 |

**End of life (subject to Japanese recycling laws; Kondo and Nakamura 2004)**

| Cost (€ per unit) | Costs (€) |
|---|---|
| 17 | 17 |

# 4 Integrating External Effects into Life Cycle Costing

*Bengt Steen, Holger Hoppe, David Hunkeler, Kerstin Lichtenvort, Wulf-Peter Schmidt, and Ernst Spindler*

**Summary**

This chapter addresses the issue of external costs. These are, specifically, non-real monetary flows that can become relevant and be monetized in the decision-relevant future or for which an economic assessment is preferred. As such, the issue of externalization has its principal bearing on societal LCC. The issue of how to account for, and possibly aggregate, a large number of social indicators (more than 200) is elaborated. The business link to sustainability, as well as the effect of a vanguard position in regard to societal and environmental behavior for firms, is presented. As societal assessment is in its infancy, and societal LCCs are few in number, this chapter, more than any other in this book, underlines the fundamentals while also proposing means to internalize costs for an idealized washing machine case.

## 4.1 INTRODUCTION

Human activities, such as business transactions and governmental decisions, have effects, which are not included in their motivation or planning. These can also influence the values of 3rd parties who are not directly involved in the business transaction or governmental decision. As such, they are external to their main goal and scope and are therefore often referred to, herein, as "externalities."

An important reason for integrating external effects in LCC is expressed in the "polluter pays principle" (PPP; Royston 1979): the polluter shall pay for the environmental damage he or she causes. The PPP has long since been an implemented principle in government policy and is also a leading principle in the EU Integrated Product Policy (IPP) process, where it has been reformulated in a less negative way as "Get the prices right" (European Union 2001, 2003b). Consequently, the costs for external effects reveal something about the potential taxes and other expenses that companies and consumers may be charged with. The extent to which different externalities need to be considered varies.

In conventional LCC, one would choose to consider externalities only if they were related to significant risks or costs. Otherwise, costs of externalities would probably be included among the "unforeseen" costs or not at all. External costs that are not immediately tangible, or borne directly by 1 of the life cycle actors in question or an immediate stakeholder, are often neglected. In environmental LCC, all environmental externalities that may turn up as real money flows (anticipated to be internalized) in the decision-relevant future (see definition in Chapter 3) would be included in a systematic way by allocating costs to environmental externalities. In societal LCC,* integrating external effects into LCC is sometimes needed. The task of a governmental organization is to consider all benefits and costs to a society that are caused by a decision, not only the direct benefits and costs (Vanclay 2003).

## 4.2 DEFINITION, IDENTIFICATION, AND CATEGORIZATION OF EXTERNALITIES

Externalities are, normally, defined as value changes caused by a business transaction though not included in its price or as side effects of economic activity (Galtung 1996). When these value changes are expressed in monetary terms, one speaks about monetized externalities. It is not obvious what to include in a list of externalities or what system to use when identifying and characterizing externalities. Different cultures and contexts may favor different assessments of externalities. Herein (see Chapter 9), the SETAC-Europe working group has chosen to link LCC as 1 of the 3 pillars of sustainable development. As such, and in anticipation of methods for social evaluation and the associated metrics, a rather broad definition of externalities for societal LCC can be anticipated. This implies, and indeed mandates, that anticipated environmental, social, and economic externalities will be included. As the main interest is LCC, it will be natural to focus on those externalities, which are possible to assign monetary values to, but others may also have an impact on the LCC for a product or service.

### 4.2.1 Selection of External Cost Categories for Inclusion

Important criteria for the identification and selection of external cost categories include the following:

- They shall fully cover all significant types of economic, environmental, and social effects due to human activities, without overlapping.
- They shall be possible to characterize in terms of category indicators, which may be understood by laypeople.
- The quantitative relation between the human activity and the impact category indicator shall be possible to model.
- It ought to be possible to estimate the monetary value of an indicator unit.

* Societal LCC is not intended to replace economic impact assessment.

Some externalities, such as excess morbidity, may have implications for environmental, social, and economic values, and the risk of double counting must be avoided, that is, impact categories covered by an LCA or societal impact assessment shall not be redundantly covered in LCC.

### 4.2.2 Categorization of Externalities

Externalities can be more or less established in the society as

- those that are already paid by someone along the value chain and are not included in the market transaction, for example municipal waste disposal, health costs, increased safety features of products beneficial for society (e.g., pedestrian protection), job security, and benefits of improved infrastructure for society. These costs would be of interest to flag in the discussion linked to a conventional or environmental LCC and would likely not be included, though this depends on the goal and scope of the case. For a societal LCC, it would be highly relevant and necessary to include them.
- those that can be monetized, are not intentionally paid, benefited, or gained by someone, and are not included in the market transaction (e.g., impacts from $CO_2$ emissions). All these costs would be of interest for societal LCC. Some of them, which could be expected to result in future costs (for instance, increased $CO_2$ tax), would be of interest for environmental LCC if it is likely or deemed to be probable in the decision-relevant future.
- those that can be monetized, are intentionally benefited by an actor, and are not included in the market transaction (e.g., free rider). Such benefits would be of interest for societal LCC and for environmental LCC only if they could be expected to be internalized in the near future.
- those that are difficult to monetize (e.g., the aesthetic value of a species or product, or wellness). In some cases, these may be of interest to societal LCC (e.g., in an interpretation phase).

Various types of externalities are identified in Sections 4.2.2 to 4.2.4 of this chapter. They represent environmental, social, or economic effects that have been observed and described in the literature. The ISO 14040/44 (2006) requires that environmental impacts on human health, ecosystems, and natural resources are considered. In SETAC-Europe's Working Group on Environmental Impact Assessment, impacts on artifacts, such as buildings and crop fields, were also mentioned (Udo de Haes et al. 2002). In social impact assessment (SIA), the impact categories discussed vary. Van Schooten et al. (2003) review current practice and suggest health and social well-being; quality of the living environment (livability); economic impacts and material well-being; cultural impacts; family and community impacts; institutional, legal, political, and equity impacts; and gender relations. Economic externalities are fairly well covered by environmental and social impact categories, though there may be others, like the ones mentioned above, and dynamic effects, which are not covered in this chapter. The methods for social externalities are evolving. There is a special working group within SETAC-Europe dealing with these issues.

Depending on the goal and scope of the LCC study, a thorough analysis may show what externalities should be covered by an LCC or by other tools (e.g., SIA and CBA). All affected and relevant impacts may then be categorized.

### 4.2.3 Consideration of 3rd Parties and Possible Sanctions

Three groups of costs — and revenues — that may be internalized in the near future have to be distinguished:

1) Costs or revenues for action (measures influencing the externality)
2) Costs that can be passed on to a 3rd party
3) Costs for sanction (i.e., for refraining from the action)

In the 1st group, those cash flows are subsumed that incur for the measure. They can be differentiated in costs and revenues for avoidance, reduction, substitution, recycling, disposal, and information and decision processes (i.e., transaction costs).

To form the 2nd group, one has to examine whether a 3rd party can and will pay the costs for environmental management. For it, the stakeholder approach might be helpful. Overall there are 4 possibilities for passing on costs: costs are passed on prospectively to customers, costs are passed on retrospectively to suppliers, costs are passed on to the government (financial aid), and finally, costs are passed on to insurance firms and, as a last possibility, a mixture of some or all aforementioned possibilities.

The difference remaining between the costs and revenues for action and the costs that can be passed on has to be compared with the costs that incur if the possible measures are not fulfilled, that is, costs like fees, penalties, and the like (group 3). Often these costs cannot be directly influenced by the company and are, therefore, referred to as "costs for sanction."

If there are several alternatives for the costs or revenues for action, the alternative with the minimum net costs or the maximum net revenues has to be selected, assuming the ecological outcome is comparable. The costs to be passed on are calculated as the sum of all feasible possibilities. If the extent of costs for sanction is not known ex ante, the expected value has to be chosen.

If the costs and revenues for action minus the costs and revenues to be passed on are less than the costs and revenues for sanction, the recommended strategy is an active one. If the situation is the other way around, the recommended strategy is a passive one, except for strategic deliberations that argue for monetary loss. Following that structure, an actor can economically manage the level of external cost he or she has to consider.

### 4.2.4 History and Ethics

For quite some time, business transactions have been made between sellers and buyers, and the price of a transaction has reflected the costs and benefits to these 2 parts. Third-party costs and benefits have only been considered when it was very obvious, like when intruding on somebody's property. A key issue for an externality to be considered is that it can be detected. The 3rd party in a transaction must know he or she is affected by it and care.

The philosopher Peter Singer (1975) has coined the term "moral circle" to describe what people care about. Various cultures have different views on what is important, and there are also large variations on the individual level. In the history of Western countries, survival of the individual and closest family members were the important things to worry about in early times. Later, the tribe and local village were included and, still later, the country, the world, future generations, animals, and so on. In addition to the mere identification of an issue in the moral circle, 2 other ways of thinking play an important role when externalities are valued (Munthe 1997), specifically, how to consider trade-offs and how to handle uncertainty. Trade-offs may be made in a common measure or by "legality." They represent 2 types of ethics: a utilitarian ethic, where things are exchangeable, and a "rights ethic," where certain conditions have to be fulfilled (Munthe 1997). The ethical dimension in economics is important, although sometimes forgotten (Sen 1987). In environmental economics and the issue of internalizing externalities, it becomes obvious and the attitude toward monetizing external environmental impacts varies.

### 4.2.5 Environmental Impacts

Environmental impacts from a life cycle perspective are described by various authors and are the subject of other SETAC working groups looking at different global, regional, and local impacts to humans or the environment. In LCA, environmental impacts are described either at the midpoint level in terms of potential impacts, such as global warming potential and acidification potential, or at the endpoint level, such as excess mortality in terms of years of lost life. As externalities are defined as value changes, impact indicators at the endpoint level are better suited to represent externalities than those at the midpoint level. They represent threats rather than value changes. For more details on impact indicators, please refer to, for example, Udo de Haes et al. (2002), Fava et al. (1993), Goedkoop and Spriesmaa (1999), and Steen (1999a, 1999b).

### 4.2.6 Social Impacts

Social impacts have, thus far, not been analyzed by SETAC. There have been some attempts to develop a social life cycle assessment (SLCA) and even to combine SLCA with environmental LCA (O'Brian et al. 1996), though the technique remains, at present, immature. However, SIA is more developed. In a recent handbook on social impact assessment, Becker and Vanclay (2003) make a distinction between biophysical or social change processes and human impacts. Biophysical change processes may be measured objectively independent of the local context. Human or social impacts are used for "impacts actually experienced by humans (at individual and higher aggregation levels) in either a corporeal (physical) or cognitive (perceptual) sense." Social change processes are demographic, economic, geographic, institutional, legal, emancipatory, empowerment, and sociocultural processes. Table A.2 in the Appendix to Chapter 4 summarizes the social impacts. This implies that one could require several hundred impact categories to make a comprehensive assessment of externalities from social impacts! Therefore, in SIA there is a screening process to sort out the most important indicators before assessing them. Chapter 9 provides some examples of approaches that reduced the number of indicators, including those that have a common baseline denominator (e.g., labor hours).

For product-related LCC studies, all these social impacts have to be related to the direct and indirect impacts of a product itself and its life cycle. This might be quite straightforward for some product features, including

- pharmaceuticals (improved actual health, perhaps balanced by unwanted side effects),
- safety features (reduced number of fatalities), and
- food packaging or refrigerators (reduced amount of food waste or better nutrition).

It is more difficult to quantify other social impacts, including improvements in

- communication, and
- individual or public mobility.

The aforementioned Table A.2 is related to quite diverse social impacts (improved relations between people, improved economics, material well being, and physical infrastructure, though also positive and negative impacts to the quality of the living environment).

### 4.2.7 External Economic Impacts

The external economic impacts are those that are not included in the market transaction (product price). Examples of such effects are given in Table 4.1 and Table 4.2.

## 4.3 MONETIZATION

There are a variety of means of defining and estimating the values of externalities or category indicators. Willingness to pay (WTP) is the amount of money a person is willing to pay for a change in his or her environment. Willingness to accept (WTA) is the amount of money a person wants to have before accepting a change. Estimations may be done by interview techniques like contingent valuation method (CVM), behavior observations, or market observations such as in hedonic pricing, where real estate prices could reveal information on environmental, economic, or social values. "Accrual basis" refers to an accounting method where expense items and income are recognized as incurred or earned, even if they have not yet been received or paid. There are legal rules for such accounting governing the recognition restricting the option of monetization of future costs to cases where the likelihood is high and the potential variance is low. Also stock prices for certain future costs are possible.

The estimation of monetary values of environmental, economic, and social impacts differs in particularly 2 aspects: discounting and who is the valuing part. Whether or not to use discounting has been dealt with in Chapter 2 for each type of LCC. Long-term effects like those of global warming, ozone depletion, and resource

**TABLE 4.1**
**General economic impacts and material well being and their relevance for LCC**

| Economic impacts | Relevance for LCC (example) | Comments |
|---|---|---|
| Economic prosperity and resilience | Societal LCC — mainly relevant for LCC studies with major investment decisions only | Could be captured by GNP changes |
| Income | Societal LCC | % change in average income of the affected regions |
| Employment | Societal LCC | % change in average employment rate of the affected regions |
| Property values | Societal LCC and sometimes conventional LCC — products or projects related to dispossession, and infrastructure projects (e.g., changing house values) | Value (change) of the affected property |
| Replacement costs of environmental and social functions (that were formerly provided by the environment, but now have to be paid for) | All types of LCC | Avoid double counting with LCA or SIA |
| Economic dependency or freedom | Conventional LCC and societal LCC — energy sector projects | Diversity of energy carriers |
| Burden of national debt (including intergenerational debts) | Societal LCC — public investment projects | Change in national debts |
| Workload or time saving or wasted time | Societal LCC and conventional LCC — many electronic products (e.g., dishwasher) | Change in workload or free time, congestion data, % of canceled or delayed trains and planes, and so on |
| Standard of living | Societal LCC — most products | |

**TABLE 4.2**
**Economic impacts from product features and their implications to LCC**

| Product feature | Relevance for LCC (example) | Comments |
|---|---|---|
| Free rider | Environmental and societal LCC — public transport (e.g., catalytic filter destroying more ozone than produced by a vehicle) and the like | Could be related to statistical data of free riders |
| Affordability | Conventional LCC — captured by product price | |
| Risks | Conventional LCC — to be captured by insurance and warranty costs | |
| Taxation | Conventional LCC — to be captured by taxation costs (differentiated for different regions) | |
| Quality or longevity | Conventional LCC — to be captured by residual value credits | |
| Future direct costs | Conventional and environmental LCC — up-coming or anticipated taxation (e.g., $CO_2$) or toll systems | Accruals or provisions |

depletion may appear insignificant or very significant depending on the discounting rate chosen.

Whose values are used for monetization is also an issue that may influence the outcome significantly. Often, for instance, European studies imply that a "western" or "northern" average citizen is the one who decides which value an impact has. The same applies to other regions of the globe. If such an average citizen uses the results as a measure of the size of an impact, this will probably be a relevant way of monetization. However, if, for instance, one is assessing measures against impacts from global warming, where the most severe effects probably will affect future generations in the arid areas in Africa and Asia, one has to be careful with whose values he or she is dealing with. Here the different values and targets clash between demands for improved development of emerging or developing economies versus demands for a conservation of current social, economic, and environmental standards in already developed countries.

A series of arguments against monetization can be found in the literature. Criticisms include the following: the approach is purely anthropocentric, is mainly western biased, increases uncertainties and injustice, relies on overly simplistic assumptions, and is only 1 of several ethical principles (see Endres 1982; Wicke 1992; Stirling 1997; Spash 1997; Schmidt 2003). This speaks for regarding monetization as 1 assessment approach that needs to be carefully judged before and when used. Any person responsible for life cycle costing, in each of its 3 forms, must therefore be careful when acting on the basis of monetary values. Although monetization is closely linked to the concept of eco-efficiency (World Business Council for Sustainable Development [WBCSD] 2003), other value concepts like "goal satisfaction" or "distance to target weighting" and comparison with what is normal may also be of interest (Steen et al. 2004).

### 4.3.1 Some Quantitative Examples of Monetized Impacts

There are several different possibilities to calculate social, environmental, and external economic impacts, and the choice of method may influence the result significantly. The further discussion is limited to 2 types that have been best analyzed so far: damage cost and prevention cost of emissions.

#### 4.3.1.1 Damage Cost

Damage cost may be defined in several ways. An often used approach is, as was discussed in the preceding section, the WTP for those affected to avoid the impact. At times, and by some, it is argued that it is not the affected people who shall "pay" but rather the polluter. Using such an argument, the WTA concept would be a better measure. However, even in cases where WTP is employed there are different ways of finding these values. For market goods such as crops, there is a market value that can be used. Depending on who is the "customer," a consumer or the state, subsidiaries may be included or excluded in the prize. If there is no direct market, methods such as CVM, where people are interviewed in a special way to find their WTP (Bateman and Willis 2002) or hedonic pricing (changes in property values), may be applied.

**TABLE 4.3**
**Damage costs from emissions resulting from different studies**

| Emission or activity | Units | ExternE (best estimate) | ExternE (low) | ExternE (high) | Pace study | Massachusetts study | EPS (2000d) |
|---|---|---|---|---|---|---|---|
| $SO_2$ | €/kg | 9.2 | 1.3 | 27 | 3.70 | 1.24 | 3.27 |
| $NO_x$ | €/kg | 10.0 | 1.1 | 30 | 1.50 | 5.38 | 2.13 |
| PM10 | €/kg | 17.0 | 1.9 | 50 | 2.17 | 3.31 | 36 |
| Cd | €/kg | 67.0 | 6.7 | 120 | — | — | 10.2 |
| Pb | €/kg | 10.0 | 5.0 | 15 | — | — | 2910 |
| $CH_4$ | €/kg | 0.2 | 0.043 | 1.6 | — | — | 2.72 |
| $CO_2$ | €/kg | 0.019 | 0.0038 | 0.139 | 0.012 | 0.018 | 0.108 |
| CO | €/kg | — | — | — | — | 0.72 | 0.331 |
| VOC | €/kg | — | — | — | — | 4.39 | 2.14 |
| Dioxine | €/kg | 290 000 | 29 000 | 520 000 | — | — | — |

Damage costs have been evaluated in the ExternE Project of the European Union (Bickel and Friedrich 2005; EU DG RTD 1995, 1999a, 1999b) and in other studies. Examples of damage costs, calculated by various of the aforementioned methods, are given in Table 4.3 (Culham 2000). Results from the ExternE project are grouped into "low," "high," and "best estimate," according to how uncertainties are dealt with. There are also possibilities to calculate different costs according to the population density around an emission source; if this population density is high, many people can be affected, and vice versa. Considering the large variations and dependence on local conditions, it may seem that these results are somewhat arbitrary and not very reliable. However, there are 2 things that compensate for this lack of precision in terms of the usefulness of the figures. One is the fact that product systems are dealt with where the processes are distributed over large areas. The other is that variations are a part of reality. A statistical approach to these data is therefore useful. Table 4.3 summarizes damage costs from emissions.

The data in Table 4.3 provide an idea of the uncertainties that are involved in estimating damage costs from emissions. Different depreciation rates, different system boundaries for the affected system, and different ways of handling uncertainty probably explain most of these differences.

#### 4.3.1.2 Prevention Cost

Herein the authors have limited the discussion to $CO_2$, the most important greenhouse gas. Damage cost would have to include damage due to the shifting of climatic zones, increase of sea level, and changed tempest and drought patterns, to name just a few, which all are subject to very high uncertainty. However, one can calculate the amount of money that is needed to prevent $CO_2$ emissions via energy-saving or efficiency-increasing activities. Table 4.4 shows some cost numbers for different $CO_2$ reduction targets (German EPA 1991). Other such studies (INFRAS/BEW 1992)

**TABLE 4.4**
**Bandwidth of cost to reduce $CO_2$ emissions in the year 2040 according to different reduction targets**

| Target of reduction of $CO_2$ emissions | 50% | 60% | 70% | 80% |
|---|---|---|---|---|
| **Cost of decreased emissions (€/t)** | 1 to 3 | 22 to 30 | 95 to 107 | 163 to 205 |

come to similar results: costs increase strongly with increased reduction targets and reach more or less the same levels.

The highest prevention cost found in the literature is some 205 €/t $CO_2$. This cost has been calculated for activities to reduce $CO_2$ emissions to 20% of today's levels (German EPA 1991). This reduction would reach the target to halve worldwide $CO_2$ emissions by reducing strongly the emissions of industrialized countries and letting "Third World" countries double their $CO_2$ emissions.

Some general points regarding prevention costs are important to note. $CO_2$ prevention cost includes much more than only $CO_2$, since preventing carbon dioxide emissions is most often realized by saving the incineration of nonrenewable resources (NRR). Under such scenarios not only $CO_2$ emissions are saved but also NRR; emissions with acute toxic, acidic, or eutrophic effects, including CO and $NO_x$; and also low-volume carcinogenic emissions such as PAH or Hg. Prevention cost is a forward-directed cost, since it does not calculate how to repair damage but how to prevent it. Prevention costs have been evaluated in studies quoted above (e.g., INFRAS/BEW 1992) and in others (Gesellschaft für umfassende Analysen GmbH [GUA] 2001a, 2001b). Data from the last study are displayed in Table 4.5. This type of cost can also vary from country to country because the labor cost for specific prevention work will vary, whereas the material cost will be quite the same in different countries. Therefore, the spread in these prevention costs will be small compared to the spread in Table 4.5.

In the assessment on future costs, one would expect that the real costs would be a mixture of damage and prevention costs. A rational behavior on a societal level would be to choose the lowest cost of the 2. In reality there are many complications related to the issue of who pays and when working against this type of rationality. Damage costs are taken by a broad population and occur later than prevention costs. For damages due to global warming and ozone depletion, the time lag between the occurrences of prevention cost and damage cost may be hundreds of years.

### 4.3.2 Monetization of Social Impacts

No systematic monetization of social impacts has been found, although CVM and hedonic pricing methods may be used for social qualities as well as for environmental qualities. A close concept is the citizen value assessment (CVA; Stolp 2003). CVA does not, however, include full monetization, only a kind of ranking, and is limited to aspects of "quality of the living environment (livability)" (see Table A.2). There are, furthermore, no principal obstacles to estimating the monetary value of social impacts (e.g., by CVM techniques). CVM is a way of determining how much people are willing to pay for goods, services, or qualities that are not commercially available.

**TABLE 4.5**
**Prevention cost for substances in air or water (€/t)**

| Substance | In the air | In the water |
|---|---|---|
| $CO_2$ biological (German EPA 1991) | 0 | — |
| $CO_2$ fossil (INFRAS/BEW 1992) | 63 | — |
| $CH_4$ (IKP/PE 2005) | 1330 | — |
| CO (IKP/PE 2005) | 76 | — |
| $SO_2$ (INFRAS/BEW 1992) | 2540 | — |
| HCl (EU DG RTD 1995) | 6100 | — |
| $NO_x$ as $NO_2$ (INFRAS/BEW 1992) | 2030 | — |
| Nonmethane volatile organic compounds (Culham 2000) | 2030 | — |
| Dust (INFRAS/BEW 1992) | 509 | — |
| CFC (Culham 2000) | 253000 | — |
| Cd (German EPA 1991; Gesellschaft für umfassende Analysen GmbH [GUA] 2001a) | 1780000 | 356000 |
| Hg (German EPA 1991; GUA 2000) | 35600 | 1781000 |
| Pb (German EPA 1991; GUA 2000) | 35600 | 71200 |
| COD (German EPA 1991) | — | 712 |
| $NH_4$ (German EPA 1991) | — | 1108 |

For practical reasons, it does not seem feasible to estimate the monetary values of 200 social changes, though it may be possible to qualitatively identify the most important ones (as in CVA) and estimate their values. Experiences from the monetization of environmental changes indicate that 3 or 4 impact types make up for almost all of the monetary values. There is also a significant body of literature that now guides the reader to identify environmental hotspots via thresholds (Rebitzer 2005). These methods are now shown to be quite valid, and one would anticipate similar possibilities for social costing over the coming decade.

## 4.4 INTERNALIZING EXTERNALITIES

Case Study Box 9 summarizes the types of damage costs that are internalized in the idealized washing machine case treated throughout this book.

Certainly some have advocated that if taxation were "fair" (i.e., socially, environmentally, and economically justified, for a given product), then sustainability assessments would be, at least for existing products, made redundant. Damage to 3rd parties shall be paid for by whoever causes it. Also the full valuation of benefits is needed for assessing the contribution to sustainable development. As an example, Table 4.6 shows what cost part this would mean for some polymer products if using maximal prevention costs as an estimate on external costs.

Materials or products with high energy demand and as a consequence a high proportion of energy cost to the overall cost are influenced significantly, as shown in Table 4.6. The cost of materials and products, which demand a high amount of handcraft, is influenced comparatively less by these maximal $CO_2$ costs as their social aspects have a high impact.

**Case Study Box 9: Externalities and Internalizing Externalities**

While carrying out the societal LCC study for the washing machine, 2 main categories of damage costs have been discussed: those related to former externalities that have already been internalized and those related to externalities anticipated to be internalized in the decision-relevant future.

The 1st type of damage costs (i.e., originally existing externalities that are already internalized) includes for the washing machine environmental services, environmental and energy coordinators, business unit environmental programs and initiatives, waste minimization and pollution prevention, fines and prosecutions, as well as environmental taxes. The latter include, depending on the geographic region, landfill, climate levy, and remediation or cleanup costs. In addition to the costs, environmental savings might be monetized as well. Examples of such savings include income savings and cost avoidance, reduced insurance from avoidance of hazardous materials and other risk reductions, reduced landfill tax as well as miscellaneous waste disposal costs, energy conservation savings, water conservation savings, reduced packaging costs, lower interest rates on loans, savings on recruiting and retaining personnel, increased revenue and/or market share, and/or increased personnel productivity.

The 2nd type of damage costs, which may be internalized in the decision-relevant future, need to be assessed by one of the methods mentioned in Chapter 4, as they are normally not covered in any business transactions, such as the following:

- Health and social well-being (e.g., less fatalities due to less emissions)
- Quality of the living environment (e.g., new gained leisure and recreation opportunities)
- Family and community impacts (e.g., less time spent for housework and more time left for the family)
- General economic impacts and material well-being (e.g., economic prosperity and reliance, and property values due to time savings)

For calculating the societal LCC of the washing machine, damage costs of emissions and resource consumptions, which have already been internalized in different studies (ExternE, Pace, Massachusetts, and EPS; Cuhlman 2000; see Table 4.3), were used. The following table gives some examples for the monetization of damage costs of main emissions related to the life cycle of the idealized washing machine. The sum of these damage costs represents the difference between the life cycle costs calculated for the environmental LCC and societal LCC, whereas externalities that have already been internalized must be included in the environmental LCC and may be included in the conventional LCC as well.

*Source*: Real case study not available; hypothetical internalized externalities based on damage cost from emissions resulting from different studies listed in Table 4.4 and life cycle inventory underlying LCIA in Table 0.1.

| Emission or resource flow | Cost (€)/kg | Cost in euros* (considered only in societal LCC) | | |
|---|---|---|---|---|
| | | Supply chain and production | Use | End of life** |
| $CO_2$ | 0.0038 to 0.139 | 15 | 104 | 0 |
| $SO_2$ | 1.3 to 27 | 41 | 72 | 0 |
| $NO_x$ | 1.1 to 30 | 4 | 27 | 0 |
| Fossil fuel | 0.025 to 0.708 | 43 | 261 | 0 |

* For consistent conventional LCC, environmental LCC, and societal LCC results' presentations, average costs are calculated. The high uncertainty of the societal LCC result is methodology intrinsic (damage costs of high uncertainty).

** End-of-life costs and savings are assumed to balance each other.

**TABLE 4.6**
**Examples of the contribution (%) of monetized ecological impacts to the estimated societal life cycle costs (sum of currently monetized and nonmonetized aspects)**

| Products made from plastics and from alternative materials | Magnitude of environmental cost (EC; % of total societal life cycle costs) |
|---|---|
| Shopping bag (polyethylene and paper) | 9.6 to 10.2 |
| Pallet (meat production) (polyethylene and wood) | 7.2 to 14.0 |
| Ham packaging (polystyrene, polypropylene, polyethyleneterephthalate, and paper) | 3.4 to 6.3 |
| Drinking water pipes (polyethylene and cast iron) | 6.5 to 10.7 |
| Thermal insulation (EPS and min. wool) | 7.6 to 11.1 |
| Windows (polyvinyl chloride [PVC], aluminium, and wood) | 3.2 to 6.2 to 9.0 |
| Floor coverings (PVC and linoleum) | 3.4 to 7.4 |
| Electrical pipes (PVC and steel) | 3.1 to 4.7 |
| E&E housings (ABS and aluminium) | 1.3 to 13.8 |
| Capacitor film (polypropylene and paper) | 2.6 to 6.7 |
| Bumper (polypropylene and steel) | 3.8 to 4.1 |
| Syringes (polypropylene and glass) | 0.5 to 1.8 |
| Infusion containers (PVC, glass) | 4.2 to 30 |

*Note:* The SETAC-Europe working group is not endorsing any of these figures; if considering uncertainties, larger ranges would be suggested.

*Source:* Based on Gesellschaft für umfassende Analysen (GUA; 2000).

A similar order of magnitude for the ratio of monetized ecological impacts to total cost, as in Table 4.6, was found using the EPS method when comparing global damage costs to global GNP (Table 4.7). The table shows that the added external costs of significant emissions and reserve depletion equal about 13% of the global GNP. To use a specific calculation related to crude oil, from which the majority of polymers are derived, the influence of maximal prevention cost on the LCC of

**TABLE 4.7**

**Weighted global emissions and resource depletions for 1990 as determined by the EPS default method***

| Substance | Global emission or reserve depletion, kg/year | EPS default index, environmental load unit (ELU)/kg | Added global value | % of adjusted global GNP |
|---|---|---|---|---|
| $CO_2$ | $2.20\ 10^{13}$ | 0.108 | $2.38\ 10^{12}$ | 2.24 |
| $SO_2$ | $1.70\ 10^{11}$ | 3.27 | $5.56\ 10^{11}$ | 0.52 |
| $NO_x$ | $1.53\ 10^{11}$ | 2.13 | $3.26\ 10^{11}$ | 0.31 |
| Fossil oil | $3.40\ 10^{12}$ | 0.506 | $1.72\ 10^{12}$ | 1.62 |
| Fossil coal | $3.17\ 10^{12}$ | 0.0498 | $1.58\ 10^{11}$ | 0.15 |
| Natural gas | $1.56\ 10^{12}$ | 1.1 | $1.72\ 10^{12}$ | 1.62 |
| Ag-ore | $1.30\ 10^{7}$ | 54 000 | $7.02\ 10^{11}$ | 0.66 |
| Al-ore | $2.11\ 10^{1}$ | 0.439 | $9.26\ 10^{9}$ | 0.01 |
| Au-ore | $1.46\ 10^{6}$ | $1.19\ 10^{6}$ | $1.74\ 10^{12}$ | 1.64 |
| Cu-ore | $9.03\ 10^{9}$ | 208 | $1.88\ 10^{12}$ | 1.77 |
| Fe-ore | $5.07\ 10^{11}$ | 0.961 | $4.87\ 10^{11}$ | 0.46 |
| Pt-ore | $1.24\ 10^{5}$ | $7.43\ 10^{6}$ | $9.21\ 10^{11}$ | 0.87 |
| Pd-ore | $9.90\ 10^{4}$ | $7.43\ 10^{6}$ | $7.36\ 10^{11}$ | 0.69 |
| Pb-ore | $2.80\ 10^{9}$ | 175 | $4.90\ 10^{11}$ | 0.46 |
| P-minerals | $1.73\ 10^{1}$ | 4.47 | $7.73\ 10^{1}$ | 0.07 |
| Total | — | — | — | 13.09 |

* Global emission and mining data from the UN and the US Geological Survey.

*Source:* Steen (1999a, 1999b).

products is high for those products that are converted quantitatively into $CO_2$, such as gasoline. Specifically, if 1 kg of gasoline is converted into some 3 kg of $CO_2$, this would increase LCC by 0.6 €/kg (approximately 0.8 €/l). One may note that the current taxation is already higher in some countries than these figures suggest.

There are however, several things preventing the PPP from being fully applied in practice, including a lack of

- knowledge of who caused what damages to whom,
- enforcement capacity,
- global and regional consensus, and
- scientific evidence for quantification of the exact damage per emission.

In the long run, these obstacles may be expected to decrease due to

- the growth of the information society, creating and disseminating knowledge on environmental causes and effects;
- increased institutionalization, resulting in bodies responsible for policy enforcement;

- globalization, promoting harmonization of language, methods, and attitudes; and
- additional research on cause–effect chains.

### 4.4.1 Sustainability's Potential Impact on Profitability and Shareholder Value

Willard (2002) has looked at 7 types of benefits of a triple bottom line for a hypothetical company, SD Inc. (Table 4.8) and found the added benefits to increase the profit on the order of 38%.

Willard's structuring of issues is different from the WBCSD's, though it contains almost the same elements (Table 4.9).

Stoeckl (2004) finds that different types of firms may benefit differently from environmental self-regulation. She mentions some key characteristics of such firms, which are

- large firms, which are likely to have comparatively low investments in relation to their turnover;
- "dirty firms," which can easily pick "the low hanging fruits";
- firms that are capable of differentiating products on environmental grounds;
- firms operating in regions of relatively high socioeconomic status or in environmentally "sensitive" areas, or dealing with environmentally "sensitive" products;
- firms selling products to relative affluent consumers;
- firms operating in highly competitive markets that have access to cost-reducing environmental programs or firms operating in very concentrated markets that have access to environmental programs that raise short-run costs and long-run benefits; and
- firms that are members of industry-wide associations.

**TABLE 4.8**
**Increase of profit in a fictive company due to applied sustainable development**

| Item | % increase of profit |
|---|---|
| Annual savings on recruiting costs | 0.03 |
| Annual savings from higher retention rates | 1.3 |
| Annual benefits on increased productivity | 25.2 |
| Annual benefits in manufacturing costs | 5.5 |
| Savings in commercial site operating costs | 0.9 |
| Increased revenue, and resulting profit | 5.0 |
| Expense reduction from reduced risks | 0.6 |
| **Total** | **38.4** |

*Source:* Willard (2002).

**TABLE 4.9**
**Comparing environmental cost and benefit issues raised by the world business council for sustainable development (WBCSD)**

| Value issues according to the WBCSD | Types of benefits according to Willard |
|---|---|
| Shareholder value | Increased revenue or market share |
| Revenue | Increasing employee productivity |
| Operational efficiency | Reduced expenses in manufacturing |
| Operational efficiency | Reduced expenses at commercial sites |
| Operational efficiency | Easier financing |
| Access to capital | Increased revenue or market share |
| Customer attraction | Increased revenue or market share |
| Brand value and reputation | Easier hiring of the best talent |
| Human and intellectual capital | Higher retention of top talent |
| Human and intellectual capital | Reduced risk |
| Risk profile | Increasing employee productivity |
| Innovation | — |
| License to operate | — |

*Source:* Heemskerk et al. (2002) and Willard (2002).

According to the WBCSD, there is a weak moderate positive relation between most sustainability issues and shareholder value and a strong relation to an environmental process focus. If Willard's estimate on the full effect of a triple bottom line, a 38% increase in profits, were correct, the shareholder value would be on the same level. However, investors likely pay more attention to the risk factors, for which Willard estimated the benefits to be on the order of a few percent. Figures of this magnitude may be found in the literature reviewed by Stoeckl (2004).

When the US Environmental Protection Agency (USEPA) published its *Toxic Release Inventory* on June 19, 1989, it led to a significant decrease in the stock prices of the company groups involved. The 1st day after the companies' names were made public, the stock prices of publicly traded firms fell on average by 0.284% (Konar and Cohen 1997). The authors refer to "the efficient market hypothesis" that "predicts that in a well-functioning capital market, security prices provide the best available unbiased estimates of the value of a company's assets."

Stock market–based measures are 1 of 2 main classes of measures of financial performance, the other being accounting-based indicators (Konar and Cohen 1997). Share price tends to be forward looking, while accounting generally reflects historic performance. Considering the complexity, and limited knowledge, regarding the relation between emissions, cost, and benefits for a company, it is unlikely that the market actors in 1989 fully understood the impact on the concerned companies' financial performance. One may also question whether the results obtained are applicable outside the United States, in countries with another legal tradition.

### 4.4.2 Revenue

Willard assumes a 5% increase of revenues due to increased prices and market share. This is mainly caused by the financial drivers, including "customer attraction" and "brand value." Stoeckl (2004 147 p) says,

> Consumers caring about the environment is a necessary — but not sufficient — condition for firm-level environmental programs to raise demands. Not only must consumers care about the environment (Condition a), but they must have access to good quality information about the environmental performance of different firms (Condition b), and they must act upon that information (Condition c).

These conditions are different for different countries and company types. An extreme impact on sales was experienced by Shell when they decided to dump the Brent Spar oil-drilling platform in the North Sea. The sales dropped more than 30% in some countries (Jensen 2002).

Stoeckl (2004) concludes that demand side effects are likely to be largest when

- firms are able to differentiate their products on environmental grounds,
- consumers care about the environment,
- consumers have access to information on environmental performance,
- consumers are wealthy and affluent, and
- firms are large.

### 4.4.3 Operational Efficiency

If revenues represent the income side, operational efficiency reflects the costs necessary to generate the sales. Willard (2002) discusses several links between sustainability issues and operational efficiency, including

- increasing employee productivity (mainly through commitment),
- reduced expenses in manufacturing, and
- reduced expenses at commercial sites.

The increase in employee productivity is partly on the individual plane and partly due to teamwork and improved working conditions. Willard (2002) estimates the benefit to increase the profits as much as 25%.

Reduced expenses in manufacturing may be due to energy savings or less material waste. Stoeckl (2004) reviews several examples on energy savings, though, of course, energy savings are easier to make when there has been little concern about this before.

Reduced expenses on commercial sites include building maintenance, temperature control, and ventilation. Energy efficiency is important here as well as for manufacturing. Other issues have to do with employee consumables, waste handling, water conservation, landscaping costs, office space, and business travel (Willard 2002).

### 4.4.4 Access to Capital

Standard & Poor (2004) use environmental criteria in rating loans with properties as security or for real estate transactions. These criteria are based on the standard ASTM E 1527-94 (American Society for Testing and Materials 1994), with some additional requirements. Their investigations include historical uses of properties in the surrounding area, hydrogeology (well records), storage tanks, polychlorinated biphenyl (PCB) items, regulatory records, environmental databases of off-site conditions, wetlands, lead-based paint, lead in drinking water, asbestos, radon, ozone-depleting substances, and compliance assessment. Environmental insurance may be used for risk management, especially in property transfer contexts. Some of the world's largest banks have carried out, for approximately 5 years now, environmental risk assessments to complement their financial ones, with the maximum "penalty" on capital of 2% per annum.

# 5 Life Cycle Costing in Life Cycle Management

*Thomas Swarr and David Hunkeler*

**Summary**

The integration of life cycle costing into existing management practices is discussed in light of the now recognized motivation for firms to track and disseminate environmental costs outside of the commercial transaction. Indicators need to be appropriately selected, complemented by the recommendation that any metrics must be based on substantiated, holistic approaches. Caution is, therefore, prescribed in regard to normalization, which can be arbitrary. The integration of environmental LCC, with or in conjunction with LCA, as a component of EcoDesign is described, along with examples as to how multinationals are bridging this issue. Communication tools, involving the presentation of LCC results, are summarized. The unique environmental management issues faced by small and medium enterprises (SMEs), as well as the barriers to compliance and risks associated, are also highlighted.

## 5.1 CORPORATE PERSPECTIVE

Firms require a management tool that is measurable and can be used to monitor internal targets as well as for external communication. Importantly, and in analogy to accounting practices, the methods and thresholds applied internally can, and likely will, differ from those used in communication with share- and stakeholders. The issue of validity is, therefore, critical, particularly in relation to externally communicated costing. Furthermore, in environmental LCC (see Chapter 3), where cost and life cycle impacts are simultaneously presented, the question of normalization can arise. Scaling the environmental impact, or the cost, for products, which one usually assesses, can render the entire analysis subjective, as a service is provided generally, even with physical goods. Therefore, enterprises, and just as much the public sector and NGOs, require methods that can be understandable, standardized, applied on a large scale, and valid, not merely for specific cases but also in a movement toward sustainability. Furthermore, if integration into a circular economy, such as that advocated by the Japanese, needs to be considered, then LCC will mandate companies to work in increasing intimacy with their partners and suppliers. While there is an interest, as it is deemed necessary, there will be limits to intercorporate interactions and financial information exchanges, and, more importantly, such transformations will take time to accept and implement, as all management concepts do.

Life cycle costing is neither financial accounting nor a means to convert indirect to direct costs, as is, for example, activity-based costing. It is also not a detailed calculation, and it must be complemented with estimates. It requires an accepted concept for data utilization and verification and must be both comprehensive and limited in scope. The concepts outlined in this introduction will be discussed, sequentially, hereafter.

## 5.2 INTEGRATING LCC INTO MANAGEMENT

### 5.2.1 Long-Term Costs

Global corporations are faced with a hypercompetitive market, which is driving the formation of extended networks of suppliers and partners and a constant push for increased efficiency and reduced costs. The global reach of industrial activity has increasingly impinged on natural systems, challenging companies to more effectively consider environmental impacts of their business decisions. These requirements drive firms to consider both monetary and physical measures of business activity. A framework for categorizing financial and physical dimensions of business decisions is shown in Figure 5.1 and provides a context for evaluating LCC objectives (Bartolomeo et al. 2000). Companies require information to support internal business decisions. They must also be able to effectively communicate information to external stakeholders to assure adequate governance systems are in place and to build market support for company investments. LCC can be used to expand organizational decision boundaries to include suppliers and customers and extend the time horizon into the future.

LCC is complementary with life cycle management (LCM), though there are important differences. LCC fits in the top 2 quadrants of Figure 5.1. However, LCM,

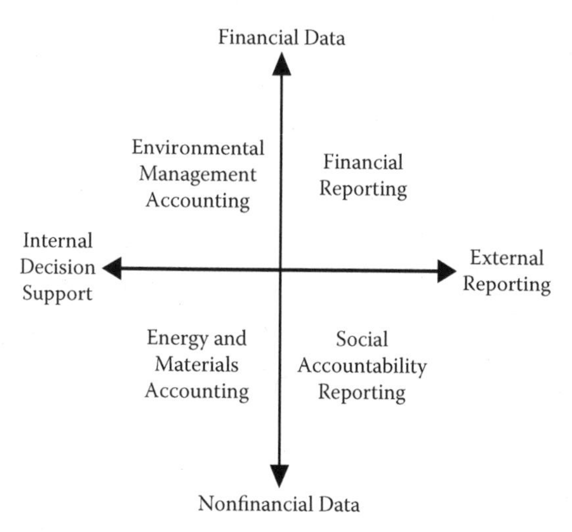


**FIGURE 5.1** Dimensions of environmental accounting, all of which can contribute to LCC.

which promotes a holistic view of the product system encompassing material and energy flows and interactions with natural systems (Hunkeler et al. 2004), is situated primarily in the left half of the figure. External stakeholders viewing the company based on voluntary and mandated reporting are represented in the right half of Figure 5.1. It can be assumed that the interest in LCC stems from a general dissatisfaction with business decisions based on conventional accounting information.

Environmental managers have been interested in LCC owing to a general consensus that existing accounting practices do not fully capture the downstream costs of many business decisions. This has been aggravated by the common practice of lumping many environmental management activities into overhead accounts that are then allocated to various cost centers. There has been significant effort in identifying various indirect or partially hidden costs, contingent costs, and less tangible image and relationship costs to estimate the true cost (USEPA 1998). These efforts are focused primarily on the cost structure that has been captured in the financial transaction (i.e., sale to a customer). Environmental LCC, therefore, tries to capture any monetary flow anticipated in the decision-relevant future, regardless of the stakeholders involved and position in the supply chain. Improved understanding and management of these costs have a direct impact on the customer's cost of ownership. This sphere of influence is shown schematically as the center element in Figure 5.2. All market transactions trigger a series of competitive and complementary actions. Some of these will address the installation, service, and upgrading of the original offering. Others may support and expand infrastructure necessary for effective utilization of the product or service. Environmental LCC can be an effective tool to identify additional business opportunities by expanding the scope of the financial analysis. These types of analyses would cover longer time horizons, though also include cost categories already addressed by the industry sector.

Figure 5.2 can also be used in reference to the 3 types of LCC identified in this book. Conventional LCC examines, generally, the commercial transaction and the industrial sector (i.e., the innermost blocks of Figure 5.2). Environmental LCC includes real costs (i.e., costs somebody is already bearing at the time of the decision) to be internalized in the decision-relevant future (i.e., "monetary external costs"). Societal LCC expands the boundaries further to include the internalization of some

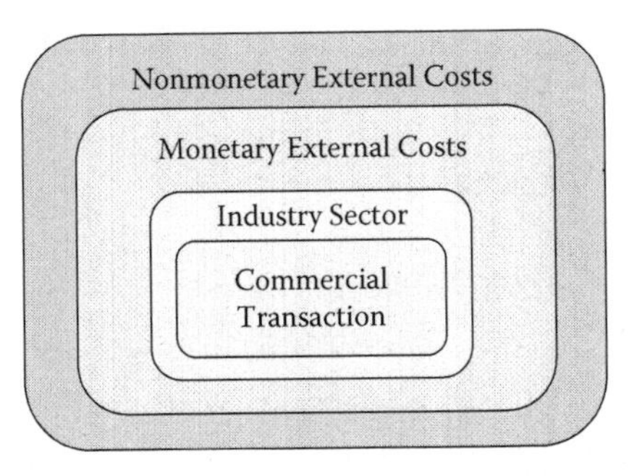


**FIGURE 5.2** Boundaries of business decisions. *Note*: Other than the nonmonetary external costs (impacts that are not considered or borne by any stakeholder in monetary terms today, i.e., at the time of decision making), all the costs listed would be included within an environmental LCC.

nonmonetary impacts (nonmonetary external costs) that could, in the long term, become relevant or monetized (e.g., the societal costs of losses of biodiversity).

An additional argument used to push for improved tracking and disclosure of environmental costs is to assist both investors and customers in avoiding potential liabilities from the indirect costs of business activity that have been shifted to people outside the commercial transaction. A good example of the shifting boundary of corporate responsibility is the EU directive on waste electrical and electronic equipment (European Union 2003a). Companies will internalize the expense of solid waste management for household electronics that was previous imposed on municipalities. Both conventional and environmental LCC can, therefore, be valuable tools for assessing future business investments or to monitor potential impacts of shifting public opinion on current business practices. These types of analyses can be used to develop innovative solutions that combine public goods into product offerings or identify those issues best addressed by government regulation or industry standards.

The aforementioned discussion is concerned with those costs that have been captured by the financial markets, through either the commercial transaction or public taxation. This obviously leads to the concern that monetary costs do not truly capture the social costs of resource consumption and environmental degradation or the area of concern shown by the lower half of Figure 5.1. A robust LCC framework will be able to link life cycle analysis studies to the monetary cost systems used by business decision makers. Unless these "dollar-driven"* decisions can also be assessed in terms of the physical limits of natural systems, it will be difficult to assess progress toward sustainability. Therefore, LCC is seen, along with life cycle assessment, as 2 of the 3 pillars in an evaluation of sustainability. The 3rd societal assessment (see Chapter 9), is still in its infancy (Hunkeler and Rebitzer 2005); it is briefly discussed in Section 5.2.5.

An example approach is the sustainability target method proposed by Lucent (Mosovsky et al. 2000). The metric attempts to gauge the amount of economic value added relative to the amount of annual carrying capacity consumed. However, it should be clearly understood that not all issues can be assessed in monetary terms. The exercise of aligning these issues with financial accounts should also be viewed as a communication tool to identify controversial issues that warrant more extensive stakeholder dialogue. Deciding which issues can be addressed through market transactions and which must be addressed by political procedures can be aided with the use of robust LCC methodologies.

### 5.2.2 Indicators and Their Normalization

Several investigations (Biswas et al. 1998; Hunkeler 1999) have noted the various types of indicators that can be considered, including microecometrics, which measure local loads often in terms of resource productivity, generally in mass or volume;

* The question of which currency one should apply in an LCC is an important issue, in particular due to the large band of fluctuations of the world's 3 main monetary units. From a corporate perspective, the issue that costs are borne in dollars, euros, and yen is accommodated using the decision-relevant future, discounting, and the currency futures market.

macroecometrics, which typically express global concentrations; and metrics, which can be used for sustainable development. Rather than discuss specific indicators, which are key to a corporation's ability to manage a situation, the issues of normalization and validation will be highlighted.

As recent cases have shown, normalization and aggregation can change even the ranking of alternatives. As an example, the selection of transition metal alloys (Park et al. 2006) differs markedly depending on how the functional unit is defined and what weighting factors are used to combine economic, environmental, and quality data. Given this, a transparent, user-independent indicator is required, and this is quite unlikely if the question of the normalization denominator can be questioned. Some single-metric scores even double normalize (Mosovsky et al. 2000). Therefore, as was shown in Section 3.3 and will be shown by the case studies in Chapter 7, the LCC working group advocates using monetary units for the LCC, with environmental measures (e.g., kg $CO_2$ equivalents) for the impact assessment. Single-score indicators are not recommended, and there is a preference that was identified by both corporate representatives and the LCC working group in general toward environmental midpoint indicators for impact assessment, rather than endpoints. It is likely that a corporate perspective on societal assessment, the aforementioned 3rd pillar of a sustainability analysis, will also prefer a midpoint, rather than single-indicator, methodology.

An example of the sensitivity of normalizing economic data with environmental impact can be understood by examining the temporal nature of taxes. While the current national, state, and local tax structures are unlikely to change, even if individual rates do, the potential of carbon taxes requires consideration. Under such a scenario of potential future taxes, external costs would be converted into internal expenses, and the same case, run several years later, would provide a higher LCC per unit of, for example, GWP. Given that ratios will be sensitive to interpretation and suffer the subjectivity of cost–benefit analysis, the SETAC LCC working group is thus not recommending any form of normalization or combination of LCC and LCIA data. Rather, the portfolio representation (see Figure 0.2 in the executive summary) of data to decision makers is advocated, as will be presented in Section 5.2.4.

### 5.2.3 Indicator Validation and Supply Chain Issues

The issue of validation is critical. Several reports (e.g., Meadows et al. 1972; von Weizsäcker et al. 1998) have been championed wherein Factor 2 to 20 reductions are recommended. However, one could reasonably question to what extent an integrated analysis has included the outsourced production, generally to SMEs, and the validity of such measures as movements toward sustainability. Furthermore, given a long-term trend to dematerialization, concomitant with a stabilization in national energy use, the issue of the exported environmental footprint becomes critical. As an example of the sensitivity of LCC to supply chain integration, one could cite that energy costs and use per unit of GDP are much higher in developing countries such as China than in the G8.* Therefore, as one changes the vertical integration of industry, in an international context, the LCC per functional unit also changes.

* Canada, France, Germany, Great Britain, Italy, Japan, Russia, and the United States.

This underlines the need for transparency not only in reporting financial data but also, more importantly, in the site of co-production and use, which should be defined from the outset. While this may seem intimidating, detailed analyses of submarket sectors are a common feature of corporate planning, though details of the target market structure are unlikely to be publicized. Therefore, given that some supply chain issues are likely to remain confidential, some publicized corporate LCC is likely to be rendered generic. Environmental LCC is, rather, seen as a tool both for external communication and in certification as well as labeling.

### 5.2.4 Presentation of LCC Results

In order to identify environmental-economic win–win situations or trade-offs, the final results of an environmental LCC study should be analyzed together with the results of the parallel LCA study. One possibility is to plot selected LCA results (e.g., 1 representative or the most important impact category as identified by the LCA interpretation) versus the LCC results ("portfolio representation," as is demonstrated in Case Study Box 10 in this subsection). One should note that if the LCA results show significant trade-offs between impact categories, or several important impact categories, then it is also possible to create several portfolios. One could also supplement the portfolio plot with a single table, as is shown in the case study box.

It is useful to note that the aforementioned portfolio presentation only shows relative differences between the alternative products studied in the combined LCA and LCC since both assessments have a comparative nature. Therefore, the resulting portfolio herein is termed "relative life cycle portfolio" so that it is not confused with the concept of Saling et al. (2002). In the future, such relative life cycle portfolios should be extended to also include the 3rd dimension of sustainability, social aspects, from a life cycle perspective.

### 5.2.5 Interfaces to Sustainable Development, IPP, and Social Aspects

Life cycle approaches have their origins in, and links to, technology assessment (see, e.g., Office of Technology Assessment [OTA] 1996), with the first studies stemming from the late 1960s and early 1970s (Hunt and Franklin 1996). These holistic approaches represent a shift from pollution prevention (see Royston 1979) and gate-to-gate concepts, which focus on single facilities of industrial enterprises, to a view that incorporates the supply chain as well as downstream processes related to a product. During this evolution, the main focus has been on methodological elaborations and building consensus on the general approaches and procedures. This important basis has led to the creation of international standards such as ISO 14040/44 (2006). The establishment of a, now quite well-accepted, LCA methodology has been possible after years of work between natural and social scientists, as well as engineers and practitioners (Marsmann 2000), and is currently being continued (Klüppel 2005). The resulting common understanding is essential for the widespread application, and one could anticipate a similar procedure for societal assessments. Indeed, the extension of the environmental life cycle view to also address economic and social aspects within sustainability seems to be needed.

## Case Study Box 10: Presentation of LCC Results

This case box summarizes the presentation of environmental LCC results. A portfolio presentation is advocated as an internal and external communication tool (see the executive summary). This portfolio plots, in the case of environmental LCC, the real monetary flows borne by any actor against the dominant impact. In the case of the idealized washing machine, the impact assessment identified the global warming potential, expressed in terms of the mass (kg) of $CO_2$ equivalents, as the dominant environmental impact. This would be reasonably expected, as the energy consumption in the use phase is the main environmental factor. As LCC must be transparent and user independent, an LCC would also indicate the distribution of costs, and for environmental LCC the impacts, across the life cycle. This is summarized in the table below.

*Source*: Real case study (consumer perspective from Rüdenauer and Grießhammer 2004) with hypothetical extensions (whole life cycle).

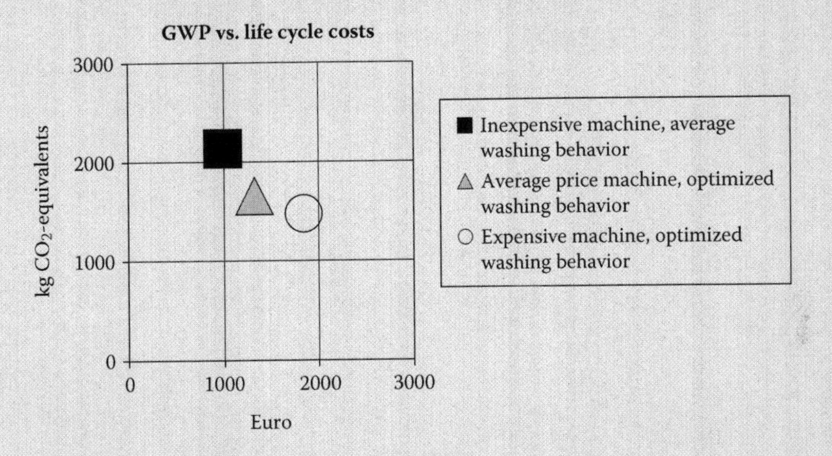


**Environmental LCC portfolio presentation of 3 alternative washing machines**

| Life cycle stage | Cost (€ per unit) | Principal impact categories | Impact (units) |
|---|---|---|---|
| R&D | 20 | Global warming | 1657 kg $CO_2$ equivalent |
| Preproduction | 216 | Acidification | 8 kg $SO_2$ equivalent |
| Production | 106 | Eutrophication | 2 kg nitrogen |
| Use | 916 | Human toxicity | 0.001 kg benzene |
| End of life (with revenues) | –42 | Resource depletion | 830 kg oil |

Sustainable development is a concept that is relatively simple to define, though difficult to quantify. Issues include the lack of metrics, as well as means to link microeconomic effects and local impacts and influences with macroeconomic or global parameters. It has also not debated how 1 stakeholder, generally a firm, is held accountable to be sustainable, at the expense of understanding why development is moving away, collectively, from community-based sustainability. A simple question could be to ask the value of a sustainable firm in a nonsustainable society. While the latter can be estimated, for example for some environmental issues, in terms of average temperatures and extreme climatic events (Allenby et al. 1998), linking product-based impacts to global parameters remains challenging. Perhaps more significantly, the assessment of sustainability will require methods for environmental, economic, and social evaluation. Although LCA seems generally accepted and is even standardized to some degree, and work on the economic dimension is presented in this book, the social methodologies remain to be formalized, as is elaborated upon in Chapter 9.

Though there had been research on social life cycle approaches and interrelations to LCA in the 1990s (O'Brian et al. 1996), this subject has not significantly advanced over the past decade. Recently, however, the social axe has regained attention, through the UNEP–SETAC Life Cycle Initiative (UNEP–SETAC 2005) and publications in journals like the *International Journal of Life Cycle Assessment* (see, e.g., Klöpffer 2003; Dreyer et al. 2006). It is clear that the assessment of the social aspects of all elements of the life cycle is a critical future issue for life cycle approaches in general. Evidence for this is for example the shift from environmental to sustainability reporting of multinational enterprises or the Millennium Goals of the United Nations (2005). Similar to the recommendations given for LCC, it seems also highly advisable to clearly define the interfaces to the environmental and economic assessments in order to build an independent dimension of sustainability (1 of 3). This independence is, in any case, a principle of sustainable development, which aims at balancing environmental, economic, and social considerations (Brundtland Commission 1987). Tendencies to methodologically integrate all impacts and benefits, whether environmental, economic, or social, into (environmental) LCA seem to be rather counterproductive in this context.

### 5.2.6 Environment and SMEs

Sustainability requires massive mobilization of human resources. It also must fit into existing structures while challenging entrenched dogma. The stress caused by incorporating new procedures, and processes, in a distribution model governed by the often changing policies of large clients is best observed through SMEs. For LCC to be complete, the collection of data from within the supply and distribution chains, in Europe approximately 70% in the hands of firms with less than 500 employees, will be critical. However, SMEs have acknowledged risks and proportionally higher overhead costs to deal with environmental health and safety issues. One key driver, therefore, is to consider environment as a direct cost, as Case Study 7.2 demonstrates.

SMEs, and new firms in particular, have, however, 1 important environmental advantage: the ability to construct, from the outset, less burdensome products and processes. This is the case because they are constructing new facilities rather than having to decide if the economics of upgrading are warranted. The higher margins for new technologies, and the lack of a need to use depreciated facilities, present an opportunity, even a competitive advantage, akin, for many firms, to their intellectual property portfolio.

## 5.3 CONTINUOUS PRODUCT IMPROVEMENT

### 5.3.1 LCC AND LCIA IN ECODESIGN

Product design has been identified as a key leverage point for promoting a shift to more sustainable business systems. LCC is a tool that can help guide the transition. Product development is essentially the evolution of information punctuated by decisions and decisions as the commitment of resources (Ullman 2001). The ultimate goal of sustainability is to redirect corporate investments toward business models that achieve a better balance of environmental protection and social equity. Thus, environmental LCC can be most effective if focused on the relevant decisions in new product development that are most likely to affect social and environmental impacts. When assessing corporate sustainability investments, the appropriate definition of "social equity" is delivering economic development to the broadest population base in accordance with the choices of a free market and democratic political system.

Research has shown many challenges to effectively integrating environmental considerations into design procedures (Handfield et al. 2001). There is confusion over what is meant by "green" products. There is a lack of accepted metrics and design methodologies. Environmental specialists speak a different language and often fail to define specific design requirements in a format that fits established practice. As it is difficult to quantify the eventual impact of specific design choices, environmental specialists often attempt to inject broad design principles and extensive data collection throughout the design process, challenging designers to find the right bit of guidance at the right time. ISO 14040/44 (2006) guidance on integrating environmental aspects into product design and development offers a representative model of the process to help proponents focus their interventions. However, there are many variants of product development processes with different lists of key tasks, and there is no one-size-fits-all solution. A review of product development research suggested that a more effective approach for integration is provided by focusing on the clusters of decisions that are highly interdependent (Krishnan and Ulrich 2001).

A simplified view of a decision-centered approach to product development is shown in Figure 5.3. Clearly, decisions made in each phase of the development process shape, and also narrow, the choices available in subsequent phases. Therefore, the environmental aspects of the production system, and indeed product, that delivers the final utility to the customer are gradually determined through this process.

Customer expectations and business goals are translated into specific, quantified design requirements that guide the integrated product development team. The key design, or functional, requirements determine part characteristics. Key part characteristics drive the design of manufacturing and product support processes. These life cycle operations eventually determine the social and environmental impacts of the product system. For example, a customer may require a certain design life to meet economic objectives. These requirements, in turn, drive the specification of parts with a certain surface hardness to provide the necessary wear resistance. The surface hardness specification determines the material selection and necessary processing to achieve those attributes. This chain of linked decisions (as is illustrated in Figure 5.3) can ultimately lead to the selection of a nitriding process that requires a cyanide copper-plating subprocess to mask the part for the nitriding operation. This would,

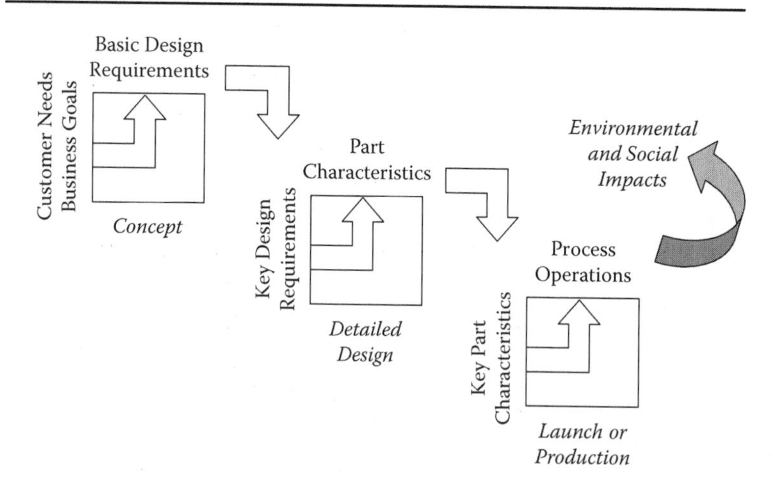


**FIGURE 5.3** Example of a decision-centered approach that is based on an established quality function deployment methodology. *Note*: See text for description.

typically, be complemented by a chromate-based stripping operation to remove the copper maskant. Thus, the generation of hazardous wastes from the aforementioned manufacturing processes is, ultimately, linked to a customer requirement for a longer design life (i.e., demand pull-based environmental impacts).

Linking the delayed and remote impacts back to specific design choices is the critical enabler for effective integration of environmental and social considerations into product development. Environmental LCC will be most effective when integrated with the businesses' decision-making processes. This model is the quality function deployment view of design, suggesting the integration challenge is not unique to environment, health, and safety objectives. Concurrent engineering is a concept that has been promoted to improve, for example, quality, producibility, or supply chain integration.

The benchmark for product development is a stage–gate product development process (Cooper 2001). A stage has a defined set of tasks that generate information, typically in the form of deliverables such as drawings, reports, and so on needed to support key business decisions. A gate is an executive business review to determine if the project should be funded through the next stage or terminated to divert limited resources to more promising projects. The review also assures that required activities have been adequately completed to support a quality decision.

A generic product design process is described in ISO TR 14062 (Margni et al. 2005). Table 5.1 provides a summary of key decisions made in the product development process. The initial planning stage surveys external pressures, public expectations, customer needs, and industry trends to define the requirements for a successful product offering. The objective is to determine the boundaries of the business opportunity and the definition of the appropriate system boundaries for environmental evaluations.

During conceptual design, the team assesses the strategic fit of the identified business opportunity with company capabilities and objectives to assure resources are focused on the most attractive projects. Detailed design activities develop a complete bill-of-material, drawings, manufacturing plans, and so on that meet technical specifications and enable design of the manufacturing and support processes consistent with project cost and quality goals. Detailed plans demonstrate project goals will be met, at least "on paper." Activities during the next stage demonstrate the feasibility of the product offering by testing prototypes or by analysis and simulation. Prescribed tasks confirm the producibility of the design and verify projected manufacturing costs. Market launch introduces the product to selected markets to validate manufacturing processes at production levels. Plans are in place to ramp up volume to meet customer demand at required levels of quality. Support systems are put in place, and product performance in the customer environment is monitored to catch any surprises. If all systems perform as expected, the project is approved for full deployment as a proven product offering. The design team works with the product management function to provide technical and logistic support to maintain the offering at warranted levels of performance. After a fixed period in service that will vary with product category and expected lifetime, a formal product review is held to assure that lessons learned from the project are captured and used to improve subsequent projects.

### 5.3.2 Evaluation Techniques, Complementary Tools, and Trade-Offs

As a summary, the corporate perspective on life cycle costing strongly advocates environmental LCC for external communication with an option for conventional LCC applied internally. This is analogous to the 2 accounting approaches applied, simultaneously, in all publicly quoted firms. Financial accounting standards, which much correspond to national or international standards and are audited, are used externally and regulated by organizations such as the Securities and Exchange Commission (SEC). Managerial accounting and a separate set of financial statements can be, and often are, used for internal decisions. Environmental LCC, which includes the economic and environmental pillars, the latter 2 from LCIA, must, in the immediate future, be complemented by rather soft estimations of social impacts to provide a sustainability perspective. As the methodologies for societal LCC evolve, in line with the development of societal assessment tools, these qualitative approaches are envisioned to be replaced by quantitative indicators.

The discussion of the differences between financial and managerial accounting, which are clearly independent concepts, can also serve as a basis to distinguish life cycle costing from these concepts. Life cycle costing does not replace any other costing approach in business and industry, but is strictly a component of sustainability assessments and has to be seen solely in the sustainability context. Therefore, care has to be taken to properly communicate the results of LCC and to make clear that these results have to be seen as complementary information in order to widen the horizon, but not as a replacement for internal or external accounting or cost estimation standards.

**TABLE 5.1**
**Inventory of product development decisions**

| Project stage | Key decisions | LCM considerations | LCC considerations |
|---|---|---|---|
| Planning | Identify business domain<br>Identify user requirements | Corporate perspective of system boundaries<br>Influences functional requirements, which drive technology and materials selection | Market allowable cost<br>Externalities to be assessed for "decision-relevant" future |
| Conceptual design | Set performance targets<br>Define system-level architecture<br>Identify alternative concepts<br>Select core technologies<br>Identify regulatory constraints | Performance drives technology and materials selection, often uses phase impacts<br>Influences EoL management, in-use support systems, and supply chain optimization<br>Complementary technologies, system lock-in<br>Public versus private goods; government affairs strategy | Cost versus performance trades<br>Initial cost versus total cost of ownership<br>Appropriate discount rates<br>Estimating uncertainties; future cost trends<br>Pricing potential liabilities in "decision-relevant" future<br>Real option value of green alternative technologies |
| Detailed design | Select concept<br>Define product architecture<br>Define manufacturing and sourcing strategy<br>Set make-buy criteria (design and/or select)<br>Define replacement parts and service strategy<br>Define quality test plan<br>Select manufacturing processes<br>Define final form and fit of parts<br>Define key characteristics<br>Set configuration of components and assembly sequence<br>Request capital appropriations<br>Define supplier selection criteria | Influences organizational and value chain structure<br>Reuse of existing components<br>EHS management versus supply chain audit; degree of control and influence<br>Reuse or remanufacture<br>Sets waste generation, emissions, and safety performance<br>Logistical impacts<br>Green message for consumers<br>Distributed service impacts | Sunk facility costs<br>Cost of supply chain surveillance; cost of supply disruption<br>Customer value proposition<br>Initial cost<br>Project profitability<br>Value of company reputation |

| Project stage | Key decisions | LCM considerations | LCC considerations |
|---|---|---|---|
| | Define marketing plan<br>Update service requirements plan<br>Define distribution plan | | |
| Testing/prototype | Establish EoL management plan<br>Establish goals for continuous improvement | EoL logistics<br>Resource recovery<br>Treatment of disposed residuals<br>Legacy system impacts | Warranty costs<br>Total cost of ownership |
| Production/launch and product review | Report metric to confirm "earned-out" processes met project goals<br>Report metrics to validate supply chain performance<br>Report metrics to monitor EoL management | Validation and assurance that system does not degrade over time<br>Impacts of fielded fleet of products<br>Data infrastructure | Incentive to replace sunk investments or cannibalize current product<br>Determining cost of value chain surveillance<br>Allocation of municipal collection and treatment costs |

The corporate perspective is aligned with a simultaneous reporting of environmental costs and impacts, without a conversion from monetary to environmental units, or vice versa. Environmental LCC comprises LCA-compatible methods, sharing, as was noted in Chapter 1, the same system boundaries. Some double counting is unavoidable, and, for this reason, the approach elucidated in Chapter 3 related to orthogonal representation of LCC and LCA is supported. Eco-efficiency methods and normalization, if not based on a systematic and validated approach, risk invalidation, and although they have been used in the past and continue to be employed, this working group does not see them as appropriate approaches for either a code of practice or a potential international standard.

As was noted earlier in this chapter, environmental LCC requires complementing the measurements and calculations inherent in LCA, and economic costing, with estimates. For the latter the field of cost estimation is well developed (see Chapter 2) in general and for particular sectors, processes, and products. For the former, LCA typically employs thresholds below which environmental burdens are assumed to be negligible, indeed 0. Although in an academic sense this may be reasonable, it will be quite difficult to justify from a corporate perspective, if for no other reason than potential liability. Hence, the business world prefers to use methods to estimate below-threshold values using management tools such as the analytical hierarchy process (AHP; Margni et al. 2005). AHP, like any estimation approach, permits one to benchmark impacts, or costs, against those for similar products. It is particularly useful in performing sensitivity analyses, which are also essential in corporate decision making.

### 5.3.3 Discussion of the Case Studies from a Corporate Perspective

The washing machine case study in Chapter 7 demonstrates several key elements of what is needed from a corporate perspective. For example, the evolution of the technology and the ageing of existing products are considered. Further, the functionality of the product represented the good as more of a service than an entity. The case also identified the key hotspot that could be used in EcoDesign: the cumulative energy demand. Some other general conclusions from the various cases, which fit well into an LCM environment, include the following:

- The portfolio representation, where LCC is plotted against a dominant impact ($CO_2$ equivalents), is a good communication tool both internally and externally (see Section 7.2 for the case on wastewater treatment).
- The sensitivity studies provide a means to interrogate the LCC and, therein, more rapidly adopt conclusions into practice (Section 7.4, carriage floor).
- The effect of discount rates *on* the decision follows typical processes that would be applied in financial assessments (Section 7.5, washing machine).
- The estimation of monetary flows likely to be internalized in the decision-relevant future provides a means of assessing potential liability and examining worst-case scenarios (Section 7.6, data transmission).
- Benchmarking of LCC by the selling price can be a useful, though qualitative, validation tool for product redesigns or the comparison of similar products.
- The EoL perspective, so important if visible consumer goods are sold, is well documented, as is an assessment of the likely costs of certification to various standards (Section 7.3, light bulbs).
- The analysis of low-volume, though high-growth-potential, niche products provides an example of a potential for significant environmental reductions as the products can be analyzed prior to being on the market and sufficiently early in their development stage to permit win–win economic-environmental savings (Section 7.1, olive oil).

Overall, the ensemble of cases in Chapter 7 can be characterized by a transparent approach, a rigor in the analysis, and an aim to uncover the key issues underlining environmental improvement possibilities. The authors of this chapter see LCC as a means to produce better products and not to perfect industry; such examples are very good guidelines to follow.

# 6 A Survey of Current Life Cycle Costing Studies

*Andreas Ciroth, Karli Verghese, and Christian Trescher*

**Summary**

A survey was distributed and statistically analyzed, at the outset of the deliberations of the SETAC-Europe working group, with the aim of identifying current practices in regard to LCC. The characterizations that were made possible based on the analysis presented herein were, therefore, the roots of the ultimate 3 types of LCC that are distinguished in this book. The 33 cases, the majority of which were performed in 2003 or later, were statistically examined in terms of their goal and scope as well as the duration of the costing, the type of sector, and the functional unit. The results are distinguished according to the geographical location of the study or production site. The means by which the various investigators examined uncertainty are discussed.

## 6.1 INTENTION

When performing LCC studies, numerous goal and scope settings are possible. These shall, ideally, be reflected in the approach and methods used in the studies, as well as in the result provided by the cases. In order to understand the current practice, a survey was distributed at the outset of the 3-year period leading up to this book. The specific goals were to identify, for the LCC in the public domain,

- different goal and scope settings, and
- various methods and methodological choices.

The correlation between the various stated objectives with the methods employed was also examined. This process could be regarded as a descriptive step. A 2nd deliverable of the survey was to make recommendations regarding the most suitable means to carry out LCC. Indeed, the survey inspired the delineation of the 3 types of LCC that are described herein and defined in Chapter 1.

It should be emphasized that the aim of this chapter is not to build a collection of cases, but rather to provide an analysis of life cycle costing studies undertaken for different use patterns, in addition to identifying possible flaws and improvement potentials for future application. In addition, the purpose was not to speak, in a representative manner, for all case studies that have been performed, nor was it to comprehensively sample existing studies. The former would be far too ambitious for the undertaking within the working group. However, the sample shall serve as a basis to formulate hypotheses, which in turn may be tested in further, more elaborated analyses. Furthermore, the guidelines established in this book on LCC, partially as a result of the aforementioned survey, have led to the selection of 7 real cases and 1 hypothetical case, as benchmarks. These are presented in Chapter 7.

## 6.2 RELATION OF THIS CHAPTER TO THE OTHER CHAPTERS

The survey was performed in 2004, and was 1 of the first activities in the LCC working group. It clearly uses a bottom-up, empirical approach (Stier 1999; Kromrey 2002), and this perspective is conserved throughout this chapter. The only exception regards the names for the different types of LCC studies, which are, for the sake of consistency throughout the book, adapted to the definitions provided in Chapter 1.

The survey presents case studies in aggregated form only. Chapter 7 will present selected case studies in more detail. In Chapter 1, the goal and scope of LCC and of LCC studies are treated in a more general, nonbottom-up perspective. Chapter 2 links goal and scope and important methodological choices for LCC studies.

## 6.3 PARAMETERS AND SETTINGS OF LCC STUDIES IN PRACTICE

There are numerous means by which to structure the different parameters for LCC studies. A survey form was developed (see Appendix to Chapter 6) to collect the key information from each case study. The survey form follows a system analysis approach, based on the following concept. When performing an LCC study, there is an object of study as well as other elements or parameters that cause a case study to occur. These, in turn, determine the result and interpretation of the case study. The effect of the person or organization carrying out the study, as well as the sources of data, measurements, and finally expert judgment or panels, should all be considered. The study processes the input data, based on methodological choices and other settings including allocation rules that "distribute" input data, as well as discount rates. Ultimately the result* that is produced contains various scenarios; itemizes the costs, often in conjunction with an impact assessment; and provides specific recommendations, while stating all assumptions. It is convenient to picture this structure in a classical box scheme (Figure 6.1).

Table 6.1 lists some possible parameters and attempts to fit them in the box scheme structure. A single parameter may have relations to more than 1 of the overarching topics (e.g., "Does the Life Cycle considered span different countries?" relates to input data, though also to the object of study, and in the end to the goal and scope).

* The term "result" may comprise quite different elements for a case study — the figure given for life cycle costs, for possibly different scenarios, being only 1 among others. The interpretation based on the findings, and also decisions based on the interpretation, are other elements that could be grouped under result.

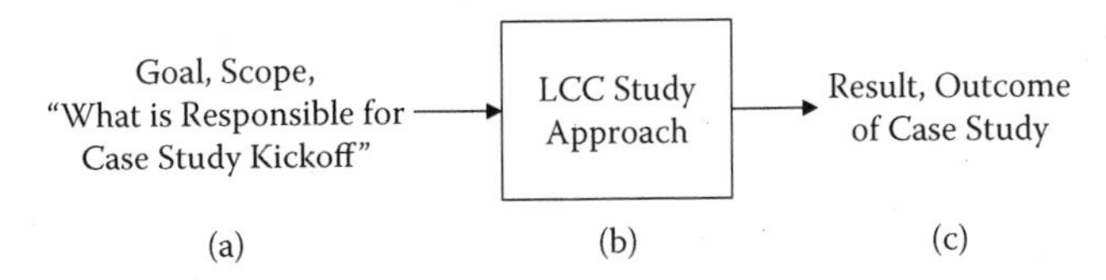


**FIGURE 6.1** Structure of an LCC study, with input, the study itself, and the result and outcome of the study.

**TABLE 6.1**
**Listing of possible parameters and settings for LCC case studies**

| Overarching topic | "Parameter"[a] |
|---|---|
| Goal and scope | Reasons for performing the study |
| | Intended use of the case |
| | Intended addressees of study |
| People | Study performed by (external contractor, internal sources, or both) |
| Object of study | Which types of branches or sectors are included? |
| | What is the object of study? |
| | Functional unit |
| | Time span covered by life cycle, per functional unit |
| Input data | Does the life cycle considered span different countries? |
| | Does it integrate costs from different sources? |
| | Quality guidelines for input data (only documented costs and prices, or also estimations of costs and prices and qualitative assessments)[b] |
| | **Case study "transaction" approach applied** |
| Approach | Source of approach (consultant, both consultant and client, or generic) |
| | Approach based on other life cycle methods (e.g., LCA) |
| | Special approaches applied (simulation, prognosis, uncertainty consideration in input data, long-term data collection) |
| | Description of different scenarios investigated, if applicable |
| | Approach of cost estimation used (price, parametric cost estimation, via functional relations, and/or other) |
| | Discounting rate as applied |
| Data sources and data processing | Data sources (company, nonpublic; market information, public statistics, and literature; and/or expert judgment) |
| | Software used (HPP [hand, pencil, and paper], spreadsheet, database, LCC or TCA tool, and/or other) |
| Other | Duration of study (initial motivation for performing the study, kickoff, and finish) |
| | Work effort required to conduct the study (person-days) |
| | **Result of the case study** |
| Costs | Overall life cycle costs per functional unit as given in study |
| | Relation of investment costs or purchasing costs to the overall life cycle costs, as given in study |
| | Type of costs, as given in result |
| | Internal costs alone or also external costs provided in result? Which type of external costs, if applicable?[c] |

*(continued)*

**TABLE 6.1**
**Listing of possible parameters and settings for LCC case studies (continued)**

| Overarching topic | "Parameter" |
|---|---|
| Explicit uncertainty in result | Uncertainty consideration in result? If yes, relative amount of uncertainty in result as given |
| | Sensitivity consideration in result? |
| Other aspects | Other aspects of object considered and investigated (reliability, energy consumption, etc.) as shown in the result |
| Life cycle | Which parts of the life cycle are excluded from the result (single life cycle stages such as production, use, maintenance, repair in use stage, recycling, and final disposal)? |
| Addressees | Addressees of study (management; client, supplier, bank, and/or others involved in companies' business; and/or public or other specific audience not involved in companies' business) |
| Interpretation and implications | What was the final interpretation of the results? Was any action taken or initiated? Is there follow-up, or are there other implications as a result of the study? |

[a] Possible realizations are specified in parentheses, if not self-evident.
[b] Not included in the survey.
[c] For a definition of internal and external costs, see Chapter 1.

## 6.4 SAMPLING PROCEDURE OF STUDIES FOR THE SURVEY

A survey was used to sample the population with the survey form, based on a preliminary version (Ciroth and Trescher 2004) as provided in the Appendix to Chapter 6. Its 3 sections correspond to the points identified in Table 6.1.

The survey forms were entered into a database as received, with some entries needing to be separated and reorganized for the analysis. As an example, lists of cost types considered were, additionally to the list provided, separated into Boolean fields indicating whether operational costs, production costs, disposal costs, and equipment and overhead costs were considered in a case study. Costs provided in studies were transformed into current euros, assuming a long-term US$–euro equivalency and disregarding the time of the study.

## 6.5 SUMMARY OF RESULTS

This section provides an overview of the statistics and calculated parameters, with complementary graphics, for the 3 types of LCC described in Chapter 1. One study in the sample performs an LCA conjoint with an LCC and assesses both internal and external costs. This study was excluded from the analysis per type of study so as to not overemphasize 1 single case. For the other external cost studies, the number in the sample is small as well; thus, results for this type of case should not be overanalyzed.

### 6.5.1 Overview of the Statistics

The survey comprises 33 studies, with most of the studies undertaken in 2003. The oldest study was conducted in 1984 (Figure 6.2). The majority of the cases were carried out in the United States and Germany, though others were from South Africa, Japan, and other European countries. One study analyzes a product over different European countries ("divers," in Figure 6.3).

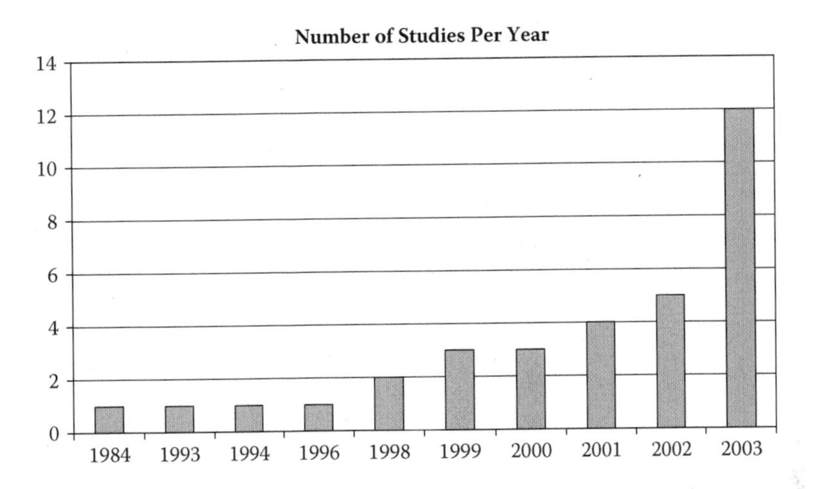


**FIGURE 6.2** LCC case studies in the survey, per year.

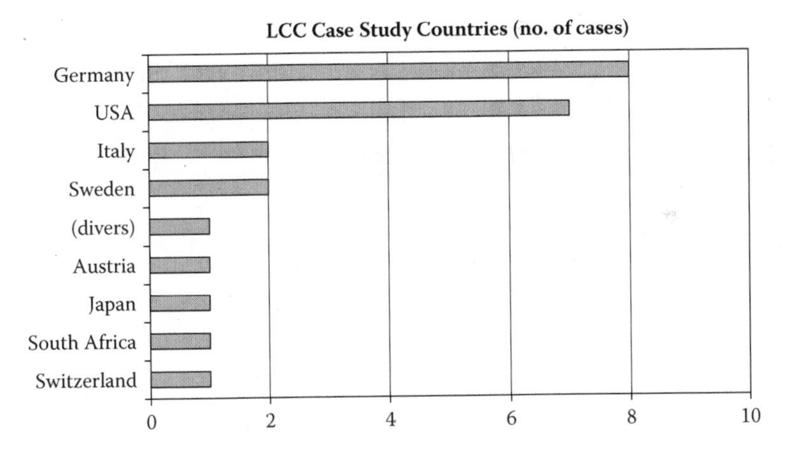


**FIGURE 6.3** LCC case studies in the survey, per country.

Figure 6.4 shows the share of the different LCC applications, as given in the survey:

1) Conventional LCC studies, using internal costs alone (>55%)
2) Societal LCC studies, using internal and external costs (10%)
3) Environmental LCC studies, using internal costs alone, in combination with LCA (>25%)
4) Societal LCC studies, using external and internal costs, in combination with LCA (<5%)*

Some studies assume a static life cycle. These were set to a life cycle duration of 0 years in Figure 6.5; the longest life cycle spans 90 years (a building).**

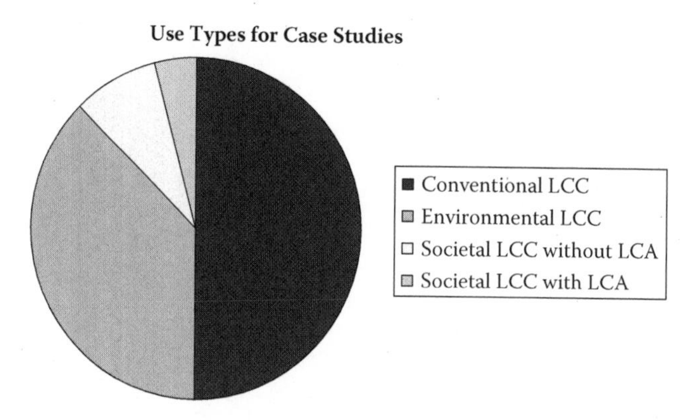


**FIGURE 6.4** Different use types of LCC studies in the survey.

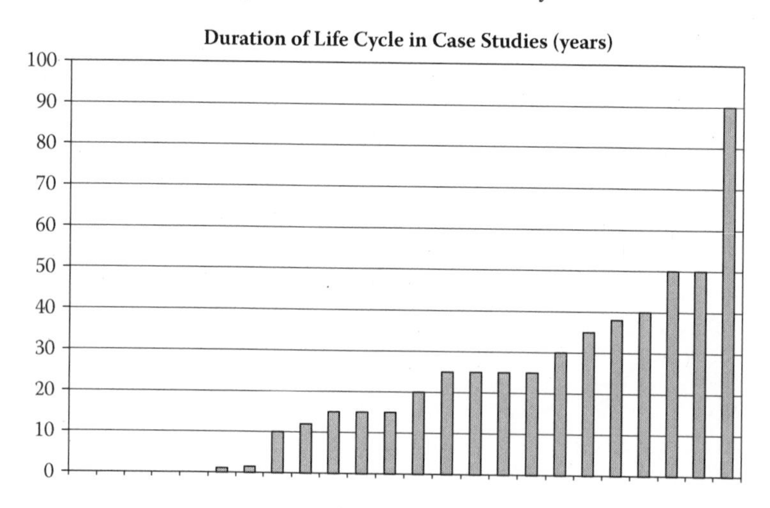


**FIGURE 6.5** Duration of life cycle in the case studies (x-axis: individual case studies).

---

* The SETAC-Europe Working Group on Life Cycle Costing recommends using a societal LCC not in combination with LCA, to avoid double counting environmental impacts.

** See Chapter 2 for a discussion on steady-state versus dynamic life cycle modeling.

Industrial sectors dealt with in the case studies are shown in Table 6.2. The numbers in the brackets indicate multiple cases. The railway and automotive industries have high shares in the case studies analyzed.

As for the industrial sectors, the objects analyzed in the studies cover a broad range from domestic boilers, to pavement design, to trams, to "solid core nuclear engines" (Table 6.3).

**TABLE 6.2**
**Industrial sectors in the case studies**

Aerospace (2)
Agriculture
Automotive (3)
Building construction and maintenance
City administration
Construction, real estate, and facility management
Domestic furnaces and boilers
Electric appliances (2)
Energy
Floor producers
Heating systems
Railway vehicles (7)
Solid waste management
Sports and leisure vehicle production
Steel industry
Street building and maintaining authorities (2)
Wastewater treatment

**TABLE 6.3**
**Objects analyzed in the case studies**

- "An average sports floor," maple-based flooring, PVC-based flooring, and poured urethane–based flooring
- Car, parts of a car, and complete car: DaimlerChrysler S-Class, and Ford Mondeo (front subframe system)
- "Chemical engines using liquid oxygen and aluminum powder [LOX/Al], solid core nuclear engines, nuclear light bulb engines, and ion engines for cis-lunar space application"
- Coupling equipment for different types of train sets
- Domestic furnaces and boilers
- Electric appliances
- Floor in a double-deck carriage (load-bearing frame, cover, finish, plywood, and aluminum structure)
- Gas-to-liquid technology to manufacture liquid fuels from natural gas
- Heat generation devices of a hospital
- Heating systems (4 different systems) to replace an existing system
- Light rail tram
- Pavement design (2)
- Power supply units for trains
- Production systems of conventional and organic extra-virgin oil
- Real estate, 4-story building in Berlin, including site and garden
- Renovation project for a prototypical data center
- Solid waste management in the Swedish municipality of Uppsala
- Stadtbahnwagen DT8 (tram)
- Tram-trains
- Two different kinds of steel materials for the production of pulp boilers: low-alloyed steel and stainless steel
- Two-wheeled personal mobility vehicles with internal gyroscopic balancing devices
- Wastewater treatment in a Swiss country town

Consequently, also the functional units in the case studies differ to a considerable extent (Table 6.4).

### 6.5.2 Costs Considered, and Not Considered, in the Case Studies

Table 6.5 lists examples of cost types as they were considered, or (explicitly) not considered, in the case studies.

The costs, as given in the form, were tentatively classified into 4 groups:

1) Production and purchase,
2) Operation and use,
3) Disposal, and
4) Equipment costs, investment costs, and overheads.

**TABLE 6.4**
**Examples of functional units in the case studies**

- "The data center undergoing renovation is a single-story structure located in a suburban community. The floor area of the data center is 40000 $ft^2$ (3,716 $m^2$). The replacement value of the data center is $20 million for the structure plus its contents."
- 1 boiler or furnace with a specific thermal performance (e.g., hot water oil boiler with 140000 Btu/hr input capacity)
- 1 $m^2$ of floor

**TABLE 6.5**
**Examples of costs considered, or explicitly not considered, in the case studies**

| Case study no. | Considered | Not considered |
|---|---|---|
| 1 | "Input costs" (also revenues, if negative): enhanced renovation capital investment, site protection capital investment, special security features capital investment, HVAC[a] upgrade capital investment future (year 17), salvage capital investment future (year …)"[b] | Not specified |
| 2 | R&D, system installation, capital costs, and operation, on the basis of a system breakdown structure | Disposal and hazards |
| 3 | Investment and maintenance costs, and repair in use stage | "Residual value costs" |
| 4 | Focus on not only environmental costs but also financial ones (investment, maintenance, energy, labor, and material)[c] | Taxes |
| 5 | Purchase/ capital costs, maintenance, operation (electricity), and others (planning, training) | Overhead costs |

[a] Heating, ventilating, and air conditioning.
[b] Note that these are citations from the forms.
[c] It seems that "environmental costs" are seen as external costs, while "financial costs" are internal costs; energy costs, for example, are then (also) part of the financial costs.

Figure 6.6 shows the share of case studies that considered these different cost types. It is clearly less common to consider end-of-life costs and overhead (OH) costs. The overall LC costs span from 10 euros to 100 million euros (Figure 6.8). They are normalized relative to the selling price of the product or functional unit, while the ratio of LCC to selling price spans from approximately 2 to 100. Some studies consider revenues and costs, which leads to negative costs (i.e., total revenues) in 1 study. Higher LC costs are typical for the "pure, internal LCC study"; see the following section. Prices are most commonly used for cost estimation (Figure 6.7). Purchase (or investment) costs vary between 0% and 85% of the total LCC, as is shown in Figure 6.9.

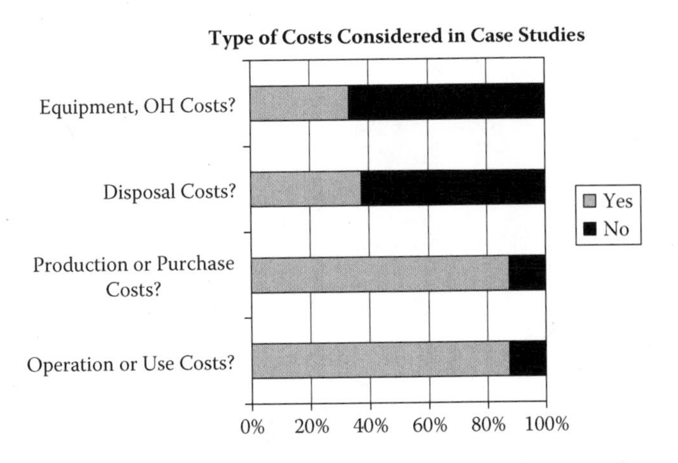


**FIGURE 6.6** Types of costs considered in case studies.

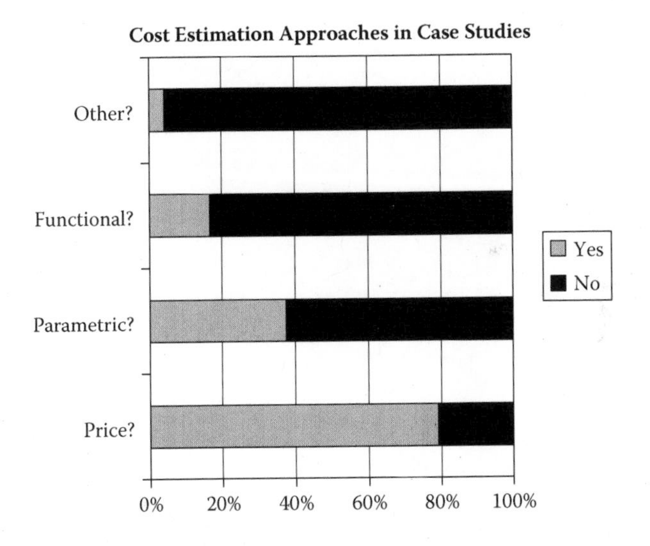


**FIGURE 6.7** Cost estimation approaches in case studies.

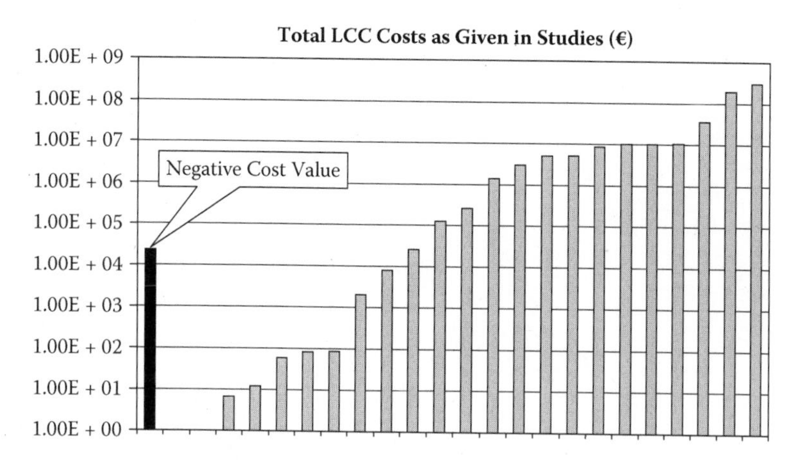


**FIGURE 6.8** Total LCC costs as given in case studies.

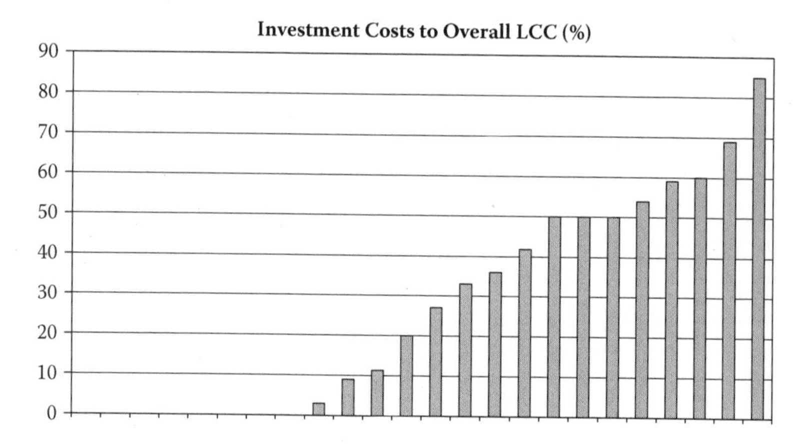


**FIGURE 6.9** Investment costs to overall life cycle costs in case studies.

### 6.5.3 Data Sources and Calculation Method Used

With respect to data sources, more than 40% of the studies claim not to use expert judgment, with most of the studies using internal and external data sources in parallel (Figure 6.10). Hand, paper, and pencil (HPP); spreadsheet software; and specialized LCC tools were each used in approximately 35% of all case studies for calculations (Figure 6.11). One should note that some LCC studies rely only on spreadsheet software and HPP.

### 6.5.4 Uncertainty and Discount Rate

The discount rate in costing has, as one aim, the consideration of uncertainty about future cash flows (see Section 2.6.1). Approximately half of the studies in the survey apply a discount rate (above 0%). A rate of 0% was assumed in cases where no rate

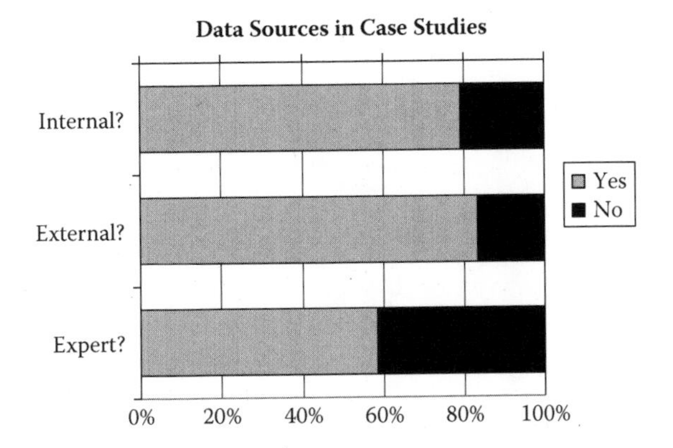


**FIGURE 6.10** Data sources in case studies.

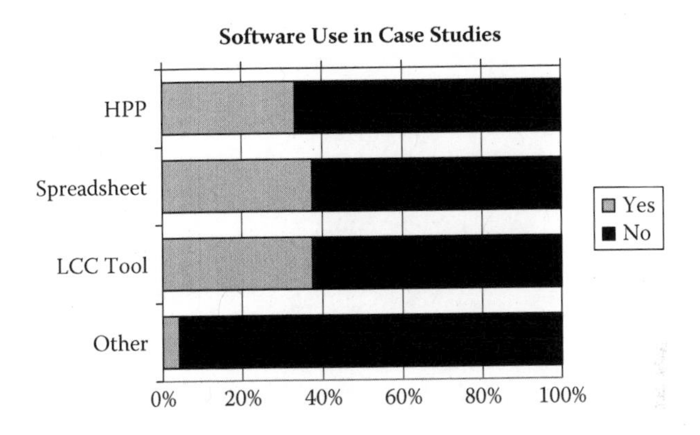


**FIGURE 6.11** Types of software used in case studies.

was reported. These cases comprise also steady-state models, where a discount rate is not applicable. A rate of 4% was applied most often (Figure 6.12) reflecting the macroeconomic reality of the cases carried out, as mentioned above, in the present (2003) low-inflation environment in Europe, North America, and Japan.

In addition, the uncertainty in today's and future cash flows, and other variants of analyses applied in the LCC studies (e.g., sensitivity analysis and Monte Carlo simulations), were considered. Specifically, the survey asked respondents whether uncertainty was considered in input data, and whether it was explicitly shown in the result of the LCC study. Few case studies in the survey show uncertainty in LCC figures explicitly, most of them being studies that consider external costs. Given the limited number of external cost studies in the survey, this finding should be handled with care. Not a single environmental LCC study in the survey provides information on uncertainty in the result.

Conventional LCC considers uncertainty in input data quite frequently (in approximately 45% of all cases). These input data are reliability and maintenance

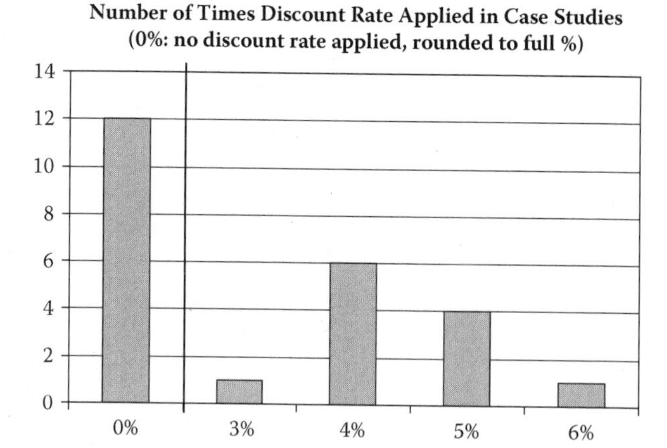


**FIGURE 6.12** Discount rate as applied in the LCC studies of the survey.

measures (e.g., mean time between failure, mean time to repair, and mean distance between failure), and also other parameters (e.g., vehicle speed, and the number of vehicles per hour, to calculate the costs of traffic jams caused by road maintenance works). It is a common approach for studies in the survey to aggregate results from simulations, and risk analysis, into single measures during the LCC calculation instead of transferring the uncertainty information to the LCC result. There is only 1 environmental LCC study in the survey that considers uncertainty in data. It is the Bahnkreis (2000) study, which analyzes railway components, making use of maintenance measures in a similar way as the "internal LCC type" studies from the railway sector. Further, only approximately 35% of all studies use prognosis techniques.* This contrasts to the fact that about 50% use a discount rate (which addresses future costs and revenues), and most of the studies analyze a life cycle of more than 10 years (see Figure 6.12 and Figure 6.13). It is apparent that environmental LCC studies in the survey do not make much use of uncertainty and prognosis techniques, as do other types of LCC application.** Long-term data collection seems to be confined to internal LCC studies in corporations.

### 6.5.5 Selected Goal and Scope, Approaches, and Result Patterns from the Survey

#### 6.5.5.1 Use Cases per Type of Application

Figure 6.14 shows use cases per type of application, with the exception of the societal LCC, which has only 1 case in the survey. Many of the studies in the survey are

* Some examples for prognosis techniques included trend analysis, regression, time series analyses, game theory, econometric or engineering models, (expert) surveys, Delphi techniques, and scenario techniques.

** This confirms a statement by Ross et al. (2002) that uncertainty in LCA is frequently discussed but not yet openly considered in LCA studies.

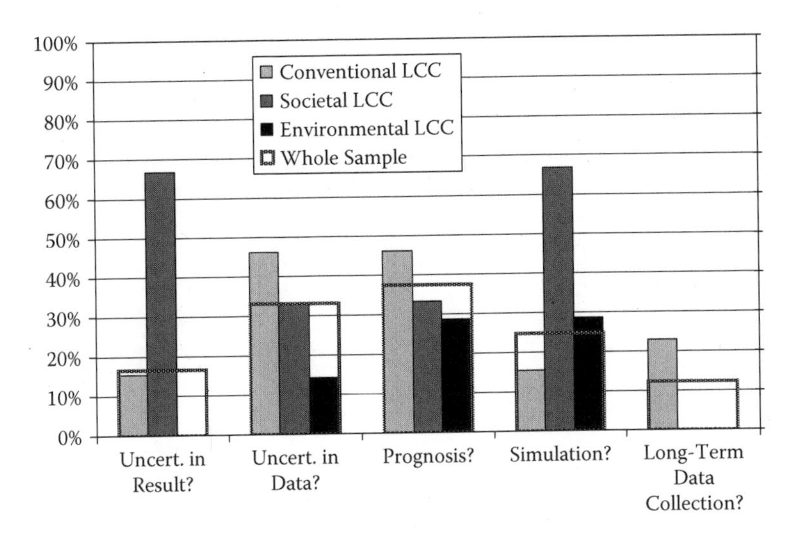


**FIGURE 6.13** Uncertainty, prognosis techniques, simulation, and a long-term data collection, for single types of LCC application and for the whole sample.

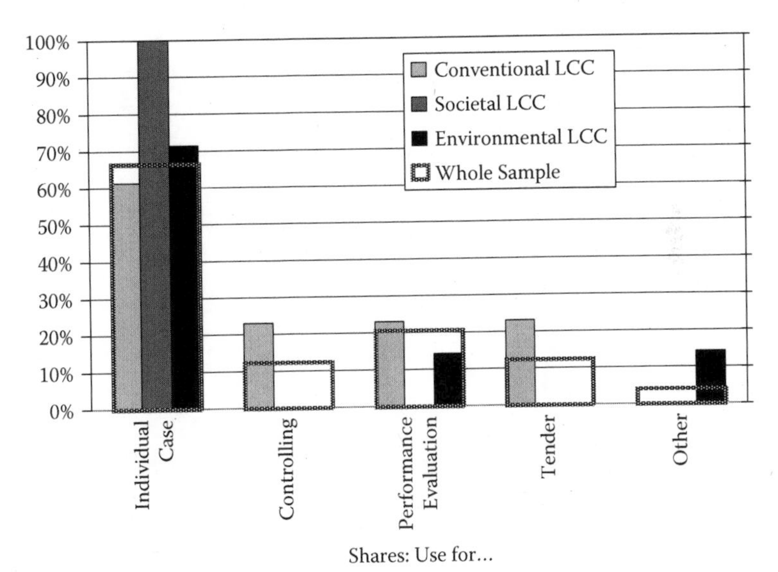


**FIGURE 6.14** Use cases, for single types of LCC application and for the whole sample (multiple entries per case possible).

individual cases, which were used to develop a method and to assess an individual case. In particular, for studies dealing with external costs, the number of cases is very small indeed, too small to draw even tentative hypotheses from. Although the internal use, performance evaluation, and tender are clear drivers, controlling still requires an explication.

### 6.5.5.2 Total Life Cycle Costs, and the Method of Cost Estimation, per Type of LC Approach

Conventional LCC studies obviously have higher overall LCC than other study types, with a "common LCC value" of more than 1 million euros and a maximum of more than 100 million euros. On the other hand, and in particular for environmental LCC, several cases with LCC of approximately 10 to 100 euros exist (Figure 6.15).* When normalizing these LCC cases by the selling price, the ratio of the LC costs to the initial payment (selling price and investment) varies between 3% and 85% (Figure 6.9), though this ratio is highly sensitive to a discount rate.**, *** This order-and-a-half variation in the ratio of LCC to selling price is noteworthy, as the return on environment concept discussed in Chapter 3 shows a similar tendency for normalized LCIA results as a function of selling price.

Distinguishing the approaches of cost estimation the studies apply, per type of study (a summary of all studies has been provided in Figures 6.6 to 6.10), demonstrates that the estimation via prices is dominant in environmental and societal LCC. This contrasts to conventional LCC, where about 40% of the cases in the survey do not use prices for cost estimation. As there were few studies assessing external costs in the sample, the figure should be treated with care for the external costing studies (Figure 6.16).

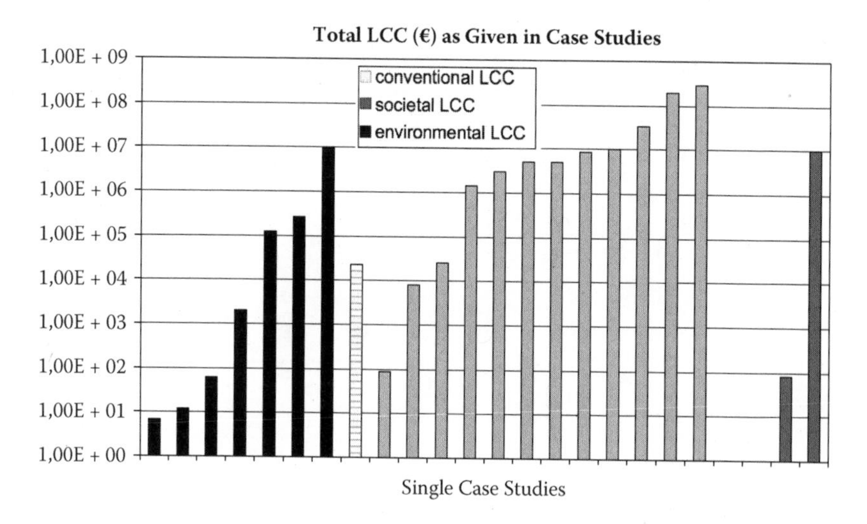


**FIGURE 6.15** Total life cycle costs, per LCC use type. Legend: Striped: negative value; x-axis: individual case studies.

* This is, of course, related to the definition of the functional unit in LCA.

** A quote from the survey in regard to the ratio of purchase costs to overall LCC noted, "50% (discounted, not discounted 10–20%)."

*** Note that the absolute costs provided by a study are not meaningless even when costs are not normalized by the selling price; in every case, costs are "normalized" by the product analyzed in the study, and by the study itself, and the amount of costs reported is a first indication of both modeling conventions and type of products analyzed in the study. It is a "first" indication because the costs are directly taken from the study without modification.

#### 6.5.5.3 Duration of Life Cycle Considered and Duration of Study by the Type of LCC

The duration of a life cycle, as considered in the studies, differs from the type of LCC study: environmental LCC studies frequently disregard time (42% of the environmental LCC studies analyzed do so), hence assuming a static, steady-state life cycle. Conventional LCC studies do not assume a stationary life cycle; in addition, the life cycle duration typically is longer (Figure 6.17). This interpretation of the figure is supported by the average, and median, of the life cycle duration. Note that the situation reverses for the duration of the studies. Although the sample is even smaller

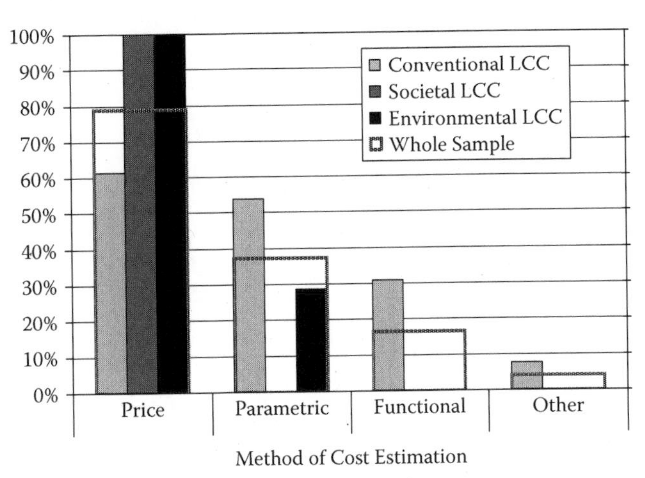


**FIGURE 6.16** Methods of cost estimation, per LCC use type (multiple entries possible).

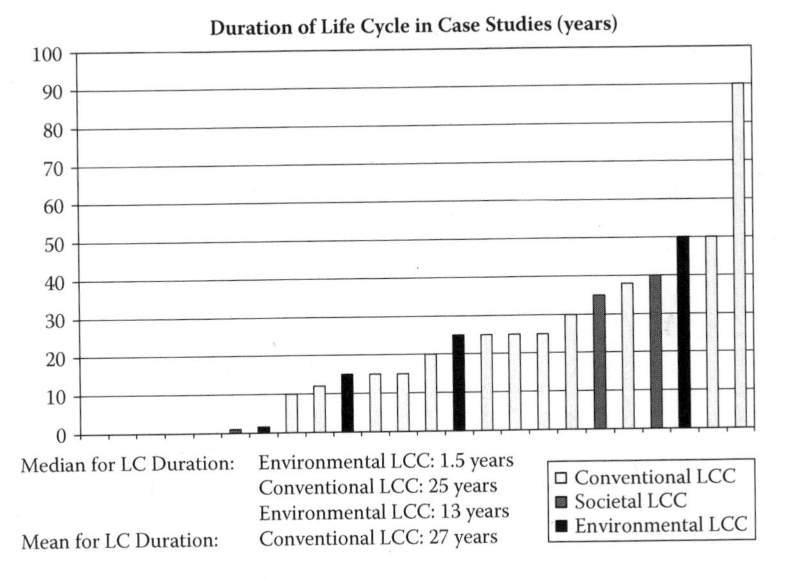


**FIGURE 6.17** Duration of life cycle in the case studies (X-axis: individual case studies).

here (it was not possible to obtain data on study duration for every study), results indicate that environmental LCC studies generally take longer (see Figure 6.18). For environmental LCC, there is a considerable difference between mean and median, caused by the many studies with a life cycle duration of 0 years.*

#### 6.5.5.4 Life Cycle Duration and LCC Discount Rate

The discount rate for most of the case studies is set to a value above 0. However, the value does not depend on the life cycle, and also studies with a long life cycle may use no discounting (i.e., a discount rate of 0), as is seen in Figure 6.19. No study in the survey discounts environmental impacts in terms of material flows or potential LCA impacts, though every societal LCC study discounts the cost.

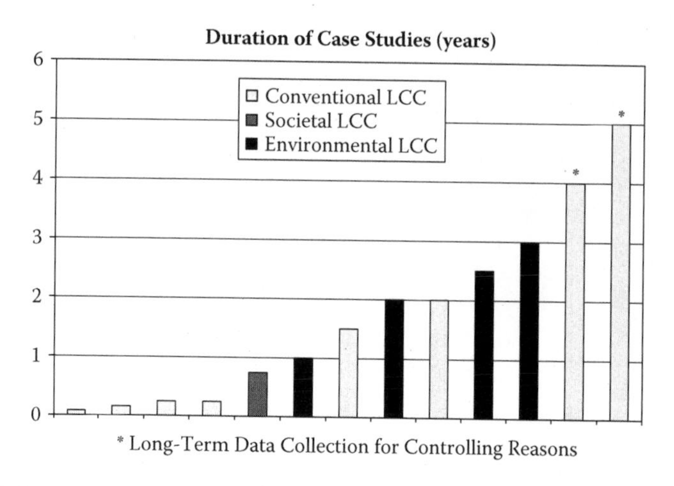


**FIGURE 6.18** Duration of LCC study for the case studies (x-axis: individual case studies).

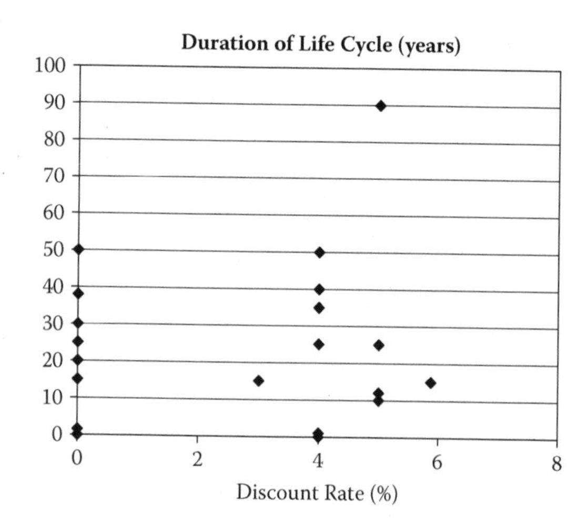


**FIGURE 6.19** Duration of life cycle plotted over discount rate.

* Life cycle duration for steady-state models was set to 0 years; see above.

#### 6.5.5.5 Addressees of the Studies per Type of Application

There seems to be a slight difference in the addressees of the different type of studies (Figure 6.20): LCC studies without a life cycle assessment (i.e., conventional LCC) are performed more for internal audiences and, to a lesser extent, for an audience along the business chain ("business to business," in Figure 6.20), while an LCC and LCA application (i.e., environmental LCC) more frequently addresses external audiences, meaning consumers, politicians, and other stakeholders not directly involved in the business of the company. For societal LCC, the number of studies in the survey is too small to draw conclusions from these results.

#### 6.5.5.6 Source of the Approach per Type of Study

The survey queried as to the source of approach. The results, given in Figure 6.21, reconfirm findings for other questions. If LCC studies are performed together with

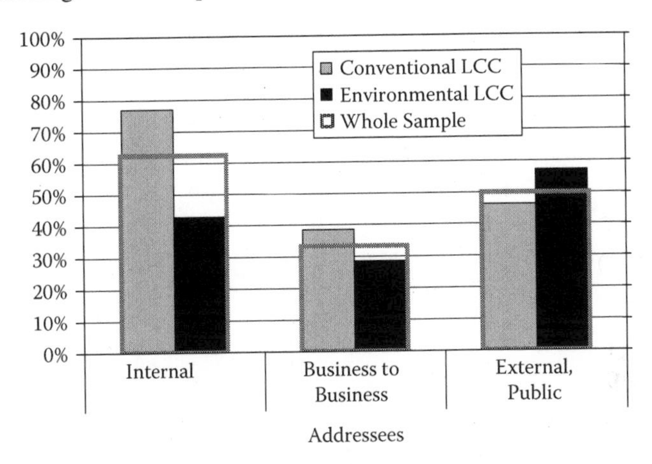


**FIGURE 6.20** Addressees of the studies, per type of study.

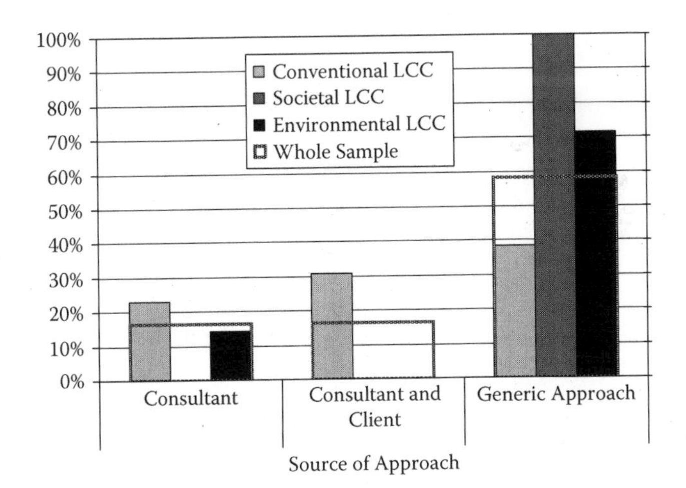


**FIGURE 6.21** Source of approach, per type of study.

LCA, or if they address external costs, a generic approach is taken. In contrast, for conventional LCC studies, it is common to use approaches proposed by the consultant or approaches that are developed conjointly by consultant and client.

## 6.6 OUTLOOK: TOWARD AN LCC CASE STUDY LIBRARY

For various application contexts, different parameter settings for the LCC case studies could be considered. An LCC practitioner might be confused regarding how to make the most appropriate choices. It is, therefore, a wise idea to have a broadly and easily accessible library, consisting of a collection of good practice (and maybe also nonrecommended practice) that guides the user toward finding good data, and the appropriate methods, for the specific problem he or she is working on. An ultimate life cycle costing code of practice could address this.

## 6.7 ANALYSIS OF THE SURVEY'S RESULTS

With its small sample size, and also with a sampling method that might be liable to biases because mainly members of the LCC working group contributed to the sample, the survey should not claim to be representative. It should not be used to support statements such as "The survey showed that …" Rather, the survey should be seen as a 1st, bottom-up, exploratory sampling leading to a hypothesis that in turn should be questioned and further analyzed later on. Its main objective was to determine the state of practice and categorize existing LCC to frame the discussion in the book. In that sense, it was successful.

According to the survey results, there seem to be several classes or clusters of LCC applications.

**Conventional LCC** is an LCC study used for internal, business-related, cost assessment, and controlling purposes. The cost assessment might consider revenues of products, as well. The product analyzed usually is complex, has a long lifetime, and has high LCC costs of up to 100 million euros. The functional unit for the product typically is "1 unit of product." The cost assessment is used in a purchase decision (tender), or from the perspective of the client, to optimize the overall costs related to a product. Costs categories, definitions, and principles of cost measurement, need to be carefully agreed upon prior to starting the assessment. Prognosis techniques, simulation techniques, and even data sampling for the product under study are utilized to better capture current and future costs and cost-influencing parameters. Overhead costs are as important as other costs, and in general are accounted for. The approach applied in the study needs to be determined by consensus between the client (who is, at the same time, decision maker) and the consultant; it is not necessarily a commonly known, or agreed on, "generic" approach. Typical examples in the literature are Seattle (2003), US Department of Transportation (1998), and Bonz (1997).

A subgroup is the "screening LCC," which designates an LCC study used for internal, business-related purposes, performed as a screening, referring to prices and expert judgment as primary data sources and, in a short time frame, using spreadsheet software as the primary software tool.

In **environmental LCC**, the product analyzed is, typically, less complex than in the conventional LCC. It is also more rigidly described, following the idea of a functional unit as given in the ISO 14040/44 ("1 m$^2$ of floor"; ISO 2006). Studies of this type are not undertaken for auditing reasons or to provide results that are input into tenders, but for investigating the environmental and economic impacts caused by a product. The studies typically follow a generally accepted, generic approach. Methods of cost estimation seem simpler than for conventional LCC studies; the studies in the sample always use prices as a way for cost estimation. Typical examples in the literature are Notarnicola et al. (2003) and Rebitzer et al. (2003). As an interpretation, the study layout is steered by LCA, and the LCA approach is broadened to also include costs.

**Societal LCC** is an LCC study with internal and external costs. A study that deals with internal costs from more than 1 perspective (costs for people in traffic jams and costs for the street authority due to street works) falls into this type, as does a study that assesses the impact of an industrial plant on the environment, and on social and economic structures in its neighborhood, in terms of costs. There have not been many studies of this type in the survey. A hypothetical example is provided in Section 7.6 on data transmission.

## 6.8 CONCLUSIONS AND QUESTIONS

Despite its moderate sample size of 33 respondents, the survey indicated use patterns of LCC studies. As indicated above, it cannot claim to be representative for any region, industrial branch, or investigation unit; it points, however, to existing deficiencies and strengths and goal, scope, and approach combinations. Due to the larger sample size, these concerns are stronger for environmental and conventional LCC. Some questions that arise are listed below, tentatively grouped into time-related, result-related, and functional unit–related groups.

### 6.8.1 Time-Related Questions

- *Steady-state modeling*: While dynamic modeling seems a natural choice for conventional LCC, especially when it deals with large investments, environmental LCC generally disregards time. This is justified as LCA, which is coupled in LCC and is regarded as a static method.
- *Prognosis techniques*: Why have they not been applied in environmental LCC thus far?
- *Uncertainty assessment*: Why is it not more broadly used in environmental LCC?

### 6.8.2 Result-Related Questions

- *Cost estimation techniques*: Why does environmental LCC rely, in the large majority, on cost estimation via prices, thereby neglecting parametric cost estimation techniques and other more advanced techniques, which are widely used in conventional LCC?

- *Control and validation* comprise an important application for conventional LCC. Will this be also a future application of environmental and societal LCC?

### 6.8.3 Functional Unit-Related Questions

- Conventional LCC tends to have larger, more complex functional units than environmental LCC. Frequently, the functional unit for conventional LCC is 1 unit of product. What are the exact reasons for this? A stricter procedure for defining the functional unit, for LCA; or a more mature method, for conventional LCC, that allows dealing with more complex products?* How should one thus define a functional unit for environmental LCC (i.e., should it reflect a product's service more so than the unit itself)?

### 6.8.4 Tentative Answers

In comparison to LCA, conventional LCC is easier to perform since the data are more readily available. They are also less abstract because the calculated life cycle costs are money units, in comparison to figures of an impact assessment. This has at least 2 consequences: it makes the result easier to understand to readers, which fuels method application, and it renders the result itself more important. In some aspects of conventional LCC application, a comparison of different LCC figures is less important than the calculated figure itself (this is certainly not the case for environmental and societal LCC). Given this, one understands better why conventional LCC aims, in these cases, to model the life cycle and the costs involved as good as possible. Parametric cost estimation techniques, dynamic models, discounting and uncertainty assessment, and, in some cases, even a monitoring of the real costs arising in a life cycle are employed. For the LCA, these more advanced techniques seem not yet necessary, thus using them may seem inefficient, although there may also be a knowledge gap to some extent.

At present, environmental LCC is LCA driven; thus, it uses primarily methods and concepts developed for LCA. This could change in the future, making LCC an equal part in the environmental LCC approach. Another development seems that LCA results may become as important as cost figures for some LCC applications today. This would then motivate the use of more advanced modeling and validation techniques in environmental LCC. However, there should be a consistent, efficient use of methods for environmental LCC, to be able to answer questions for economic and environmental sustainability in an efficient, valid, and easy-to-communicate manner (see Chapter 7).

To those questions where tentative answers are not provided, some will be addressed in the recommendations in Chapter 8, as, for example, sensitivity analyses

* To some extent, the smaller functional unit for LCA is a mere linear scaling problem: for example, whether for a floor the functional unit is "1 $m^2$" or "the whole floor of 65 $m^2$" does not make an analysis more complex. However, in other cases (e.g. "one typical insurance building with 4500 PCs and 9 floors," defined for a conventional LCC), it does not seem possible to split this unit into smaller units that are more "accessible" for LCA.

seem critical in methods that will be paired with, and need to mesh with, others. Other questions, like the use of prognosis techniques, will evolve in LCC as a science, perhaps eventually modifying a code of practice, though unlikely a part of it from the outset.

# 7 Life Cycle Costing Case Studies

*Andreas Ciroth, Carl-Otto Gensch, Edeltraud Günther, Holger Hoppe, David Hunkeler, Gjalt Huppes, Kerstin Lichtenvort, Kjerstin Ludvig, Bruno Notarnicola, Andrea Pelzeter, Martina Prox, Gerald Rebitzer, Ina Rüdenauer, and Karli Verghese*

**Summary**

Examples are provided of environmental and conventional LCC for both durable and nondurable goods, as well as services. Common conventional LCC still dominates the real case studies, with a few environmental LCC examples. As no complete societal LCC was identified in the literature, a hypothetical application related to data transmission is presented. The cases are intended to serve as references as to how LCC results should be presented, the methodology that is appropriate, and the level of documentation required. Products with different market lives are discussed, with the technology spectrum varying from food to high-tech electronics developments.

## 7.1 INTRODUCTION

Various studies are summarized that provide examples for conventional, environmental, and societal life cycle costing. They are intended to describe the methodology and provide specific examples of the data required, calculations, validation, and presentation of the results. The cases include examples of durable, semidurable, and nondurable goods, with product lifetimes ranging from months to decades. High-tech and commodity examples are included, identifying cases where various materials of choice (EcoDesign), downstream burdens (e.g., transport and disposal), and process variations dominate the impact. Examples are generally based on real data from the private sector, with 2 cases presented from the consumer perspective. There is also 1 hypothetical case included to demonstrate the societal LCC methodology.

The case studies presented are as follows:

- Organic versus conventional extra-virgin olive oil (Section 7.2)
- Wastewater treatment (Section 7.3)
- Energy-saving lamps versus incandescent lamps (light bulbs; Section 7.4)

- Different construction variants for a double-deck carriage floor, a component of a regional train (Section 7.5)
- Washing machines (Section 7.6)
- Hypothetical case on data transmission (Section 7.7)
- A consumer perspective of the utilization of an automobile (Section 7.8)
- Residential buildings, including both static and dynamic evaluations (Section 7.9)

Table 7.1 characterizes the 7 cases based on real data. A variety of product lifetimes are considered, from those that are consumed in months (olive oil) to those with a durability measured in decades (e.g., water treatment and residential buildings). The cases also include technology-related products such as family transport (automobiles) and data transport, those for which the use phase is critical (i.e., light bulbs and washing machines), as well as services (transport carriage).

All case studies are summarized in Table 7.2, with the overall life cycle cost expressed in monetary units (euros) as well as the key environmental impacts identified in the studies, as far as possible. In each subsection, detailed discussions of the individual cases will be presented in a common format. Cases are presented where maintenance dominates (train carriage) as well as others where the transport phase greatly exceeds all other costs, and impacts, for a service (water treatment). Some of the studies relied on very detailed engineering models and simulations, whereas others were LCIA-based for which supplemental LCCs were added. As Table 7.2 demonstrates, the ratio of the LCC to the selling price can differ significantly (from a factor of 2 to more than 1000) depending on if the product use phase is important to the overall operating costs. Interestingly, for the automobile, where the use phase dominates the environmental impact, it accounts for only 50% of the life cycle cost. For buildings, however, where construction is a major impact, the use phase is more than 90% of the total cost. This implies that, for LCC to be normalized or benchmarked, it must done within a very homogeneous product group.

Table 7.1 and Table 7.2 reveal that case studies having been carried out in practice are still predominantly applying conventional LCC (4 conventional LCC case studies versus 2 environmental LCC and 1 societal LCC case study). The fact that environmental LCC would add value to many of the conventional studies carried out

**TABLE 7.1**
**Characterization of the life cycle costing case studies evaluated**

| Sector of activity | Case studies evaluated | Geographical region | Type of life cycle costing |
|---|---|---|---|
| Manufacturing: durable goods | Train carriage | Europe (Germany) | Environmental LCC |
| | Light bulbs | Europe (Germany) | Conventional LCC |
| | Washing machines | Europe (Germany) | Conventional LCC |
| | Automobiles | Europe (Germany) | Conventional LCC |
| Manufacturing: nondurables | Olive oil | Europe (Italy) | Societal LCC |
| Service | Water treatment | Europe (Switzerland) | Environmental LCC |
| | Residential building | Europe (Germany) | Conventional LCC |

is a valuable justification for the new method presented in this book. For example, the light bulb case study considers only the energy consumption in the use phase, admittedly the most important environmental impact. However, an environmental LCC comprising a complete LCA would underpin the pros and cons of the currently discussed phasing out of incandescent lamps in Australia and Europe; for example, assessing properly the mercury used in the alternative compact fluorescent lamps (CFL) and internalizing $CO_2$ costs from emission trading. A societal LCC would even assess the societal implications of a shift of incandescent lamp factories currently located in Europe to CFL factories in Asia.

Environmental LCC would require an assessment of the end-of-life phase in the washing machine case study, which may put into perspective a too early substitution of a less energy-efficient washing machine. It would require a complete LCA of the passenger cars investigated in the automobile case study, which may prove, if the estimation of environmental impacts of an entire car reflects the environmental impacts accurately, an important issue in the increasing public rating of cars in particular by NGOs. Environmental LCC would also require sophisticated calculation of the energy consumption of a building, based on U-values of different building elements, in relation to LCIA indicators like global warming potential (GWP), ozone depletion potential (ODP), nitrification potential (NP), eutrophication potential (EP), or photochemical ozone creation potential (POCP) over the whole life cycle of a building, which may differentiate the results of the building case study dependent on age, climate zone, and annual energy consumption per $m^2$. For all conventional LCC case studies, it would be of interest to learn about the implication of the environmental costs under discussion, which are likely to become mandatory for the manufacturer in the decision-relevant future. These would include $CO_2$ costs from emission trading, $CO_2$ taxes or binding targets for cars, minimum energy performance standards (MEPS) for appliances, and compliance costs with legislation like the European Environmental Performance of Buildings directive (European Union 2005a).

The 2 environmental LCC case studies, water treatment and train carriage, both lead to airtight (and quite likely very nonintuitive) conclusions after having studied all economic and environmental impacts over the whole life cycle. These are that the transport of water treatment sludge to ultimate disposal dominates the environmental impacts for distances above 40 km and that maintenance accounts for 75% of train carriage LCC, whereas energy in use only sums up to 16%.

The olive oil case study demonstrates well the current state of the art of societal LCC; in fact, key external costs are considered according to the path-breaking Extern-E methodology (Bickel and Friedrich 2005). However, more societal impacts have not been considered in the available real case studies. As a comprehensive example on societal LCC, considering mainly the government and society perspective, a hypothetical high-tech case study on data transmission will be presented in Section 7.6. This case study considers subsidies and VAT and internalizes all environmental damages, including those for which there are no real money flows (yet) for data transmission companies. Even this hypothetical societal LCC could better incorporate qualitative societal impacts, as outlined in Chapter 4 (e.g., standard of living, employment, and working hours).

The cases selected for presentation were those that the working group, following the 3 years of deliberations, felt would pass review for an international standard should, for example, ISO develop 1 for LCC in analogy to ISO 14040/44 (2006) defined for LCA.

**TABLE 7.2**
**Summary of life cycle costing case studies**

| Case study | Life cycle cost (€ per unit) | Selling price (€ per unit) | Life cycle assessment principal impacts | Type of LCC | Comments |
|---|---|---|---|---|---|
| Olive oil—organic and traditional | Organic: 5680 € (internal costs); traditional: 3796 € (internal costs)<br>Organic: 1103 € (external costs); traditional: 10 403 € (external costs) | N/A | Considered pesticide and fertilizer use, agricultural activities on water, transport, energy, and packaging | Societal LCC, key external costs considered | Extern-E project |
| Water treatment | $120 per person per year (30% solids, 100 km transport)<br>$80 per person per year (25% solids, 40 km transport) | — | Transport of sludge to ultimate disposal dominates the impacts for distances above 40 km | Environmental LCC | Transport dominates environmental impact and LCC |
| Light bulbs | Energy-saving type 1: 1808.68 €<br>Energy-saving type 2: 3595.06 € | Energy-saving type 1: 15.45 €<br>Energy-saving type 2: 7.60 € | Use phase (impacts not assessed in traditional LCC) | Conventional LCC | The inclusion of costs for $CO_2$ would now be easily possible because of the European emission-trading scheme, which |

| | | | | | |
|---|---|---|---|---|---|
| | Incandescent lamp: 5614.51 € | Incandescent lamp: 1.20 € | | | allocates a price to the emission caused |
| Train carriage | 248 000 € (purchase: 3%; maintenance: 75%; and energy in use: 16%) | 7440 € | Use phase, and energy related to transport | Environmental LCC | Maintenance accounts for 75% of LCC |
| Washing machine | 1168 € (purchase: 43%; energy supply: 22%; and water supply: 35%) | 500 € | Not identified (only energy for production and direct energy consumption in use phase) | Conventional LCC | — |
| Automobile | Corsa 1.0: 10 945 €<br>Punto 1.2: 10 890 €<br>Citroën C2: 10 990 € | Corsa 1.0: 19 964 €<br>Punto 1.2: 2116 €<br>Citroën C2: 19 119 € | An overall measure of environmental impact was estimated using the VCD methodology | Conventional LCC | An environmental assessment is included, though as the system boundaries differ from the LCC, the assessment remains "conventional" rather than "environmental" |
| Building | Residential: 134 471 €<br>Mixed-use: 1 465 994 € | Residential: 2 854 340 €<br>Mixed-use: 17 813 206 € | No LCA carried out | Conventional LCC | Inclusion of the time value of money as a scenario |

## 7.2 ORGANIC VERSUS CONVENTIONAL EXTRA-VIRGIN OLIVE OIL

### 7.2.1 Summary

Organic olive oil production in Italy has grown in recent years, presently covering 1.2 million hectares (ha), though it still remains a niche product. The production systems for conventional and organic extra-virgin olive oil were compared, in order to assess their environmental and cost profiles, and to verify if the 2 dimensions, environmental and economic, converge in the same direction (Notarnicola et al. 2003). This case presents an example of a societal LCC, though it is incomplete as only key external costs are considered.

### 7.2.2 Definition of the Case Study

Olive oil production in Puglia, a region of the south of Italy, represents 50% of the entire Italian production and 18% of the EU production output. In recent years, the production of organic extra-virgin olive oil has increased due to new consumer behavior and to the high organoleptic, nutritional, and healthiness qualities of this product. The total Italian "organic" growing area is approximately 1 200 000 ha, featuring more than 60 000 farms. However, organic extra-virgin olive oil still remains a niche product because of its higher market price than other oils and fats, and due to the cost of labor in the extremely delicate operation of olive harvesting and the additional costs due to the minor yields (about 30%) of the organic soil. The functional unit was the conventional and organic production of 1 kg extra-virgin olive oil (cradle-to-gate analysis). The internal and external costs are respectively shown in Table 7.3 and Table 7.4.

### 7.2.3 Entry Gate and Drivers

Various olivicultures and olive oil producers, both conventional and organic, have been involved in supplying data and should be viewed as the entry gates. The higher cost of the olive oil (both conventional and organic) compared to other oils and fats was the driver for change.

### 7.2.4 Implementation

#### Barriers

There have been problems due to the use of fertilizer and pesticide diffusion models, and enhanced scientific support to predict their fate in the environment is needed.

#### Process to Achieve Change

A rationalization of the use of fertilizers and pesticides could lead to a reduction in the external costs in the olive oil life cycle. In regard to the internal costs, the labor in the agricultural phase is the most relevant.

**TABLE 7.3**
**Internal costs of organic and conventional extra-virgin olive oil production per functional unit (€)**

| Cost item | Organic | Conventional |
|---|---|---|
| | **Agricultural phase** | |
| Pesticides | 0.171 | 0.117 |
| Fertilizers | 0.268 | 0.181 |
| Lube oil | 0.023 | 0.011 |
| Electric energy | 0.143 | 0.085 |
| Water | 0.077 | 0.046 |
| Diesel | 0.084 | 0.048 |
| Labor | 4.344 | 2.864 |
| Organic certification costs | 0.064 | — |
| **Total (agricultural phase)** | **5.174** | **3.352** |
| | **Transport phase** | |
| Transport | **0.078** | **0.039** |
| | **Industrial phase** | |
| Electric energy | 0.014 | 0.024 |
| Labor | 0.089 | 0.045 |
| Water | 0.002 | 0.022 |
| Packaging | 0.298 | 0.298 |
| Waste authority | 0.015 | 0.015 |
| Organic certification costs | 0.009 | — |
| HACCP certification costs | 0.0009 | 0.0009 |
| **Total (industrial phase)** | **0.428** | **0.405** |
| Total | 5.680 | 3.796 |

*Source:* Notarnicola et al. (2003).

**TABLE 7.4**
**External costs of organic and conventional extra-virgin olive oil production per functional unit (€)**

| Cost item | Organic | Conventional |
|---|---|---|
| External costs of energy | 0.664 | 0.533 |
| External costs of fertilizers and pesticides | 0.439 | 9.870 |

*Source:* Notarnicola et al. (2003).

## Successes, Results, and Benefits

Detailed environmental and cost inventories of the 2 olive oils have been carried out and disseminated.

## General Learnings

Figure 7.1 shows the differences between including and excluding external costs in the LCC. If one does not consider the external costs, the organic oil has a higher cost profile that is due to its lower agricultural yields. However, when external costs and less tangible, hidden, and indirect costs are included, this results in the organic oil having a lower total cost compared to the conventional oil. This result illustrates the need to account for external costs, as has recently been initiated by the European Commission. The options for environmental improvement in the conventional system are, primarily, related to a more reasonable use of pesticides while, in the case of the organic system, a reuse of the brushwood as fuel, rather than their uncontrolled burning on the fields, which could lead to a better environmental profile both in the human toxicity (HT) and in the photochemical ozone creation (POCP). Moreover, in the organic system the traditional extraction method has been used in the inventory setup. It should be noted that the Associazione Italiana per l'Agricoltura Biologica (AIAB) guidelines (2007) permit organic oil producers to apply the "continuous-extraction method," which is characterized by energy consumption double that of the traditional process. It would be desirable to note, in these guidelines, the relevance of energy consumption, since the consumer who is interested in organic foods would like to buy a more ecocompatible product, which is characterized not only by the absence of chemical fertilizers and pesticides but also by an overall environmental advantage.

### 7.2.5 Overview of Tools Used

The environmental LCC methodology used was based upon the guidelines stated by White et al. (1996), which divide the costs into 3 categories: conventional corporate costs (typical costs that appear in the company accounts); less tangible, hidden, and indirect costs (less measurable and quantifiable, often obscured by placement in an overheads account); and external costs (the costs that are not paid by the polluter, but by the polluted). The physical and economic data were collected directly from farms, olive oil factories, and public databases, as will be highlighted below.

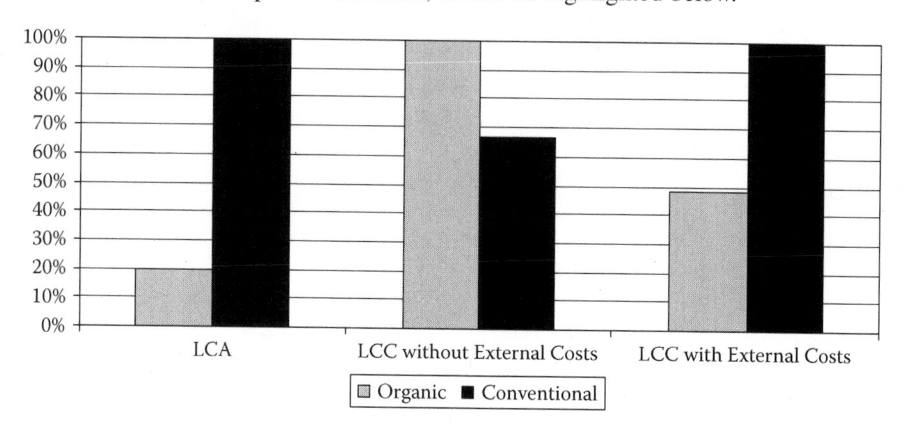


**FIGURE 7.1** LCA–LCC with and without external costs for conventional and organic extra-virgin olive oil production. *Source*: Notarnicola et al. (2003).

The external costs relative to the energy have been taken from the ExternE National Implementation Italian Report (FEEM 1997), while those relative to the use of pesticides and fertilizers were taken from a study of the Bocconi, Milan, Italy, in which the production and social costs of organic and conventional agriculture have been compared. The study took into account the impact of the agricultural activities on the water and monetized these impacts, showing that the damage caused by conventional agriculture due to fertilizers and pesticides in terms of reclamation and decontamination costs is 33 times higher than that caused by organic agriculture. The Department of Commodity Science, Faculty of Economics, Bari, undertook the study.

## 7.3 WASTEWATER TREATMENT

### 7.3.1 Summary

An environmental life cycle costing study of municipal wastewater treatment in Switzerland was undertaken, with the results being directly applicable also to other European countries. It was found that the inclusion of both upstream and downstream processes is essential for determining improved options for wastewater treatment.

### 7.3.2 Definition of the Case Study

When assessing options for the treatment of municipal wastewater and supporting decision making in this context, one must focus not only on the quality of the end product, the cleaned water, but also on the costs for the operation of the wastewater treatment plant. The impacts and costs caused by the operation of the plant as well as by upstream processes (e.g., the production of ancillaries) and downstream operations (e.g., treatment and transport of produced sludge) also need to be taken into account. The aim of this case study was to analyze both environmental impacts and costs of the complete life cycle of wastewater treatment, in order to identify the drivers for environmental impacts and costs, to identify trade-offs, and to give recommendations for improved and more sustainable wastewater management. A detailed elaboration of the case study is given by Rebitzer et al. (2003). The study examined medium-sized (50000 person equivalents) municipal wastewater treatments, with biological treatment followed by sludge digestion.

In this study typical municipal wastewater treatment options in Switzerland were assessed, with the general findings being transferable to other European countries. The complete system of wastewater treatment was examined, taking all involved processes into account as illustrated in Figure 7.2.

The reference flow, which was identical with the functional unit in LCA terms, to be assessed was the treatment of the average amount of a typical municipal wastewater per year and person in Switzerland. The perspective of a company or municipality operating the wastewater treatment plant is chosen because these are the organizations concerned with the costs of the treatment and associated processes. Additionally, and even more importantly, these organizations can influence the system of wastewater treatment.

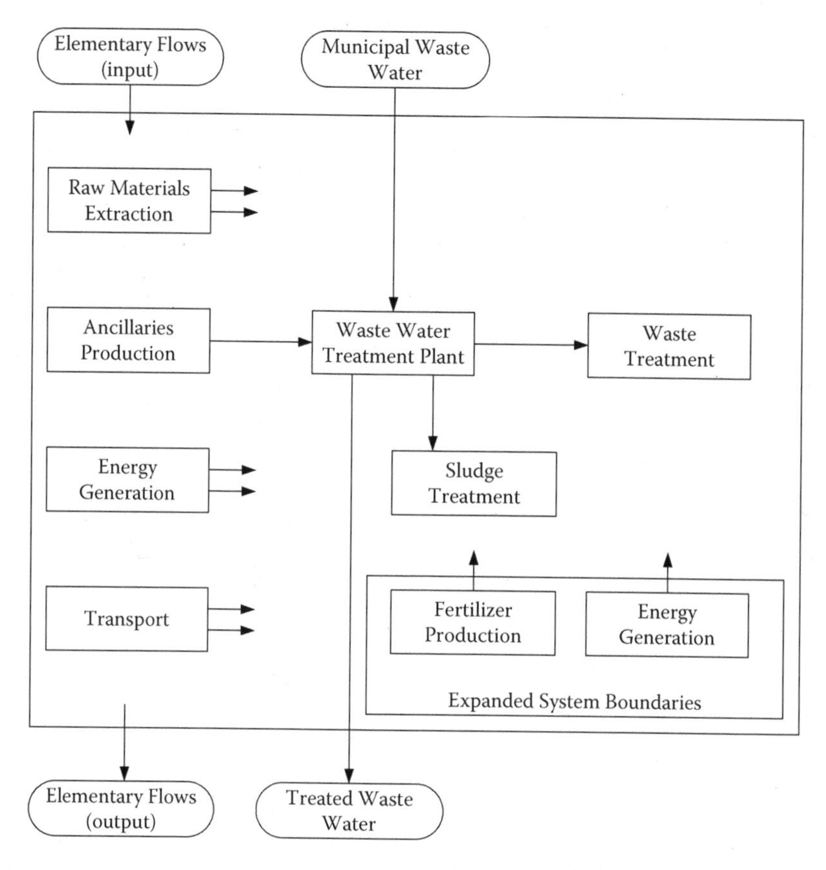


**FIGURE 7.2** Model of the LCA system for municipal wastewater treatment. *Source*: Rebitzer et al. (2002).

The results were used to create a basis for the planning of new wastewater treatment plants (in the sense of design for environment), as well as to assist decisions in existing plants for the treatment of municipal wastewater. The different options (scenarios and assumptions) assessed are listed in Table 7.5.

The methodology of life cycle inventory–based LCC was employed for this case study, where the life cycle cost assessment is based on the life cycle inventory of an LCA and where both LCC and LCA are separately considered for decision making (for a detailed presentation of this approach, see Chapter 3 of this book). In this specific case, since no long-term intervals are involved, discounting was not applied. The results of the different options were elaborated in detail, also analyzing the contributions of single elements of the system (see Figure 7.3 as an example for 1 scenario) and the most important parameters (Figure 7.4).

The case study demonstrates that dry substance of the sludge and transport distance are extremely important parameters, which can lead to differences in variable costs up to a factor of 3 (Figure 7.3). The additional costs for advanced flocculants for achieving a higher dry content are very small in relation to the cost savings that occur downstream. If the results of the LCA are compared (see Rebitzer et al.

**TABLE 7.5**
**Studied wastewater treatment scenarios and assumptions for the treatment of typical municipal wastewater in Switzerland**

| | Scenario A | Scenario B | Scenario C |
|---|---|---|---|
| Inorganic chemical for phosphorous removal (coagulation) | — | Iron sulphate | Iron sulphate |
| Organic chemical for sludge dewatering (flocculation) | — | — | Cationic polyacrylamides |
| Specification of wastewater treatment plant | 10 000- to 50 000-person equivalents | 10 000- to 50 000-person equivalents, adapted to aforementioned chemical use | 10 000- to 50 000-person equivalents, adapted to aforementioned chemical use |
| Sludge disposal | Incineration or agriculture | Incineration or agriculture | Incineration or agriculture |
| Transport distances for sludge disposal | 40,100, and 200 km | 40,100, and 200 km | 40,100, and 200 km |

*Source:* Rebitzer et al. (2003).

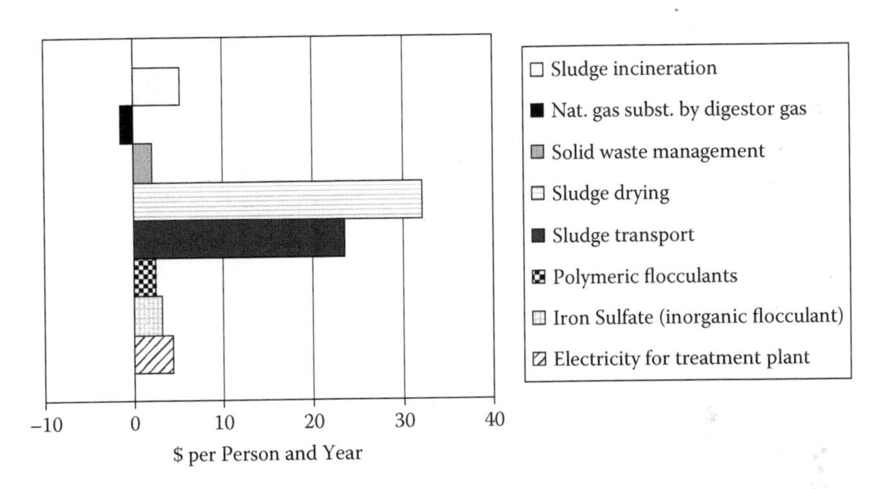


**FIGURE 7.3** Costs of the different elements of the system of wastewater treatment (scenario C, with incineration of sludge). *Note*: Assuming 40 km transport distance and sludge with a dry content of 35% leaving the wastewater treatment plant. *Source*: Rebitzer et al. (2003).

2003), the same parameters also highly influence the environmental impacts, leading to a comparable ranking of the options. Therefore, the use of advanced (highly soluble, generating higher sludge dry material) flocculants can be seen as an environmental-economic win–win situation and an important contribution to the goals of sustainable development. Overall, it is the environmental cost, as well as burden,

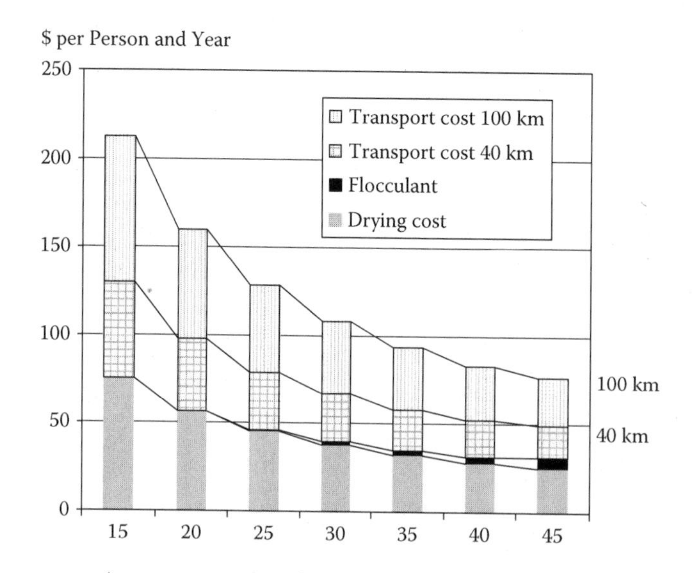


**FIGURE 7.4** Significant variable costs of municipal wastewater treatment as a function of sludge dry substance and disposal transport distance. *Source*: Rebitzer et al. (2003).

of the truck transport of high-water-containing residues that dominates. Reducing the water level (i.e., via better chemical drying) is essential to improve the environmental cost or impact. A "green" flocculant is not one that is made from a naturally extracted macromolecule but one that can provide highly dry material and reduce the transport burden.

Both the environmental analysis and the life cycle costing study were carried out by the Life Cycle Systems group of the Swiss Federal Institute of Technology Lausanne (http://www.epfl.ch; Ecole Polytechnique Fédérale de Lausanne 2007) in close cooperation with the firm AQUA+TECH (http://www.aquaplustech.com; AQUA+TECH n.d.), a developer and provider of flocculants for wastewater treatment and other applications.

### 7.3.3 Entry Gate and Drivers

The entry gate of the firm (AQUA+TECH) was through the top management of the company. The director initiated the study, delivered the required information, and supported any internal staff necessary for the study.

In the process industries, which deal with wastewater treatment, as well as in municipalities and other operators of municipal wastewater treatment plants, there is high cost pressure and the need for efficient solutions that fulfill the requirements of environmental regulations. On the other hand, the costs are often only addressed for single elements of the life cycle of water treatment. Specifically, many operators of treatment plants try to optimize their internal costs without looking at the downstream implications. Therefore, the study was driven by the aim to make the overall costs as well as the interactions between the different elements of the system transparent in order to raise awareness and to gain understanding for economically and environmentally improved options. If operators realize what a difference they can make in regard to the downstream operations of sludge transport and disposal, one can create incentives for sharing small additional costs that yield much greater overall savings. In this context, the study was also driven by the requirement to use such results for the sales and marketing of advanced* flocculants.

### 7.3.4 Implementation

The main barrier for the implementation of the findings was that, often, different actors control the different elements of the life cycle, and each actor (e.g., operator of water treatment plant and companies, or municipalities running the sludge transport and disposal business) tries to optimize its own costs and revenues, which often does not lead to an overall optimum. There may, for example, be conflicting interests if reduced transport costs lead to a decrease of revenue for the transport business. This is exacerbated by the accounting practices of some communities, where water treatment and sludge transport are in separate budgets. Such conflicts of interest sometimes even occur if the different processes are part of 1 organization, specifically if they have to operate as profit centers. Other barriers include the perceived communization of the market, which implies that clients, often municipalities, seek to purchase inexpensive product, on a per kg basis, rather than economically effective solutions (e.g., euro per ton of water treated).

The key to achieve changes and enable the implementation of life cycle thinking is awareness rising and education for all actors involved. In addition, the targeted communication of the results of the life cycle costing study is essential. Further steps could include the organization of round tables and supplier–customer interaction in the sense of supply chain management (see Slagmulder 2002).

The case study shows that, using the life cycle approach, environmental-economic win–win situations can be identified. For the first time, the benefits of flocculants could be demonstrated from a systems perspective. This can be seen as a central step in moving wastewater treatment to a more sustainable practice, in spite of the barriers for implementation that still exist (see above). However, as many municipalities separate the chemical budget from transport in different cost centers, life cycle thinking, while understood in European water treatment, has been slow in integrating into purchase decisions.

In addition to the aforementioned learnings, this case demonstrated that life cycle costing, if based on the life cycle inventory of an LCA, is an easy-to-apply and efficient approach for assessing the economic dimension of sustainability. From a life cycle management point of view, one can conclude that the "reuse" of LCA data for LCC is a promising way to better integrate life cycle thinking into decision making in industry and other organizations.

* Historically, a flocculant that could provide higher dry material levels for the cake had a dosage, and cost, penalty. A newer generation of synthetic materials, entering the market in approximately the year 2000, overcame this disadvantage, basically with improved solubility. These "advanced" materials are so named as they permit the simultaneous improvement of the cost and environmental attributes related to water treatment.

### 7.3.5 Overview of Tools Used

The tools that were applied in this case study were life cycle assessment methodology according to ISO 14040/44 (2006), life cycle inventory–based life cycle costing according to Rebitzer (2005; see also Chapter 3 of this book), and physical-chemical process modeling of wastewater treatment as presented in Braune (2002).

## 7.4 A Comparison of Energy Saving and Incandescent Light Bulbs

### 7.4.1 Summary

This case study compares 3 types of bulbs (i.e., 2 different types of energy-saving lamps and 1 incandescent lamp) using conventional LCC in combination with a qualitative analysis of the ecological aspects (Günther and Kriegbaum 1999). The data for the production of lamps are directly supplied from the producer. The case is presented from the customer's perspective.

### 7.4.2 Definition of the Case Study

In order to identify the best solution for lighting a room, the conventional LCC method has been applied. As a basis for the calculation, an office with 50 sockets for light bulbs with a maximum of 75W is chosen. All sockets were assumed to be employed. The case study was conducted by a large German lamp producer in cooperation with academics in order to compare different types of lamps and to show the economic superiority of energy-saving lamps.

In the analysis, 2 different types of energy-saving lamps (one high-price, high-tech product and one low-price, low-tech product) are compared to a traditional incandescent lamp. For data concerning the different types of lamps, see Table 7.6. Furthermore, the duration of use per day and the days of use per month have to be determined in order to calculate the use time in months or years. In the case study, 12 hours of use per day and 21 workdays per month were applied as bases.

In addition to the information provided above, the company supplied further documentation concerning the amount of $CO_2$ produced in the production process of

**TABLE 7.6**
**Life cycle costing data for 3 alternative bulbs**

| | | Energy-saving lamps | |
|---|---|---|---|
| **Alternative** | **Incandescent lamp** | **Type 1** | **Type 2** |
| Acquisition costs | 1.20 € | 15.45 € | 7.60 € |
| Life span | 1000 h | 10000 h | 3000 h |
| Energy consumption | 75 watt | 15 watt | 30 watt |
| Energy costs | 0.225 €/kWh | 0.225 €/kWh | 0.225 €/kWh |
| Disposal costs | None | 0.625 € | 0.625 € |

the lamps. Therefore, in combination with the energy consumption data, an environmental evaluation on the basis of $CO_2$ was also possible.

A conventional LCC, from the customer's perspective, was the method of choice. The costs included are acquisition price, energy consumption costs, and disposal costs. The inflation is considered with a rate of 2%, and a discounting rate based on the cost of capital is used (4.7% per annum). Moreover, time (e.g., useful life) and performance (e.g., energy consumption) have to be considered as well. All economic results are assessed using sensitivity analysis. The ecological information (energy consumption for production and use, the related $CO_2$ equivalents, the amount of waste produced, and hazardous materials) were assessed in a qualitative way.

### 7.4.3 Entry Gate and Drivers

The objective of the study was to assess the economic benefits of the use of energy-saving lamps as seen by the purchaser. Therefore, the results of the study should promote the use of high-quality energy-saving lamps. Data for the case study were provided from internal company sources (e.g., life span) and publicly available sources (e.g., energy prices).

Normally, conventional LCC is used for expensive products or projects. This case study shows the usefulness and impact of the LCC results even for day-to-day utility-driven "services." The impact of 1 lamp seems minor, though if one considers the number of lamps used, the impact becomes large. In addition, the identification of all relevant measures for the decision is important. Therefore, a combination with an ecological assessment could be carried out to identify the true relevance of the decision even if this is not the scope of conventional LCC. It should be noted that this case, given its limited internalization of externalities, and the reduced system boundary are neither, respectively, a societal nor an environmental LCC.

### 7.4.4 Implementation

A problem arises from the existence of different prices for energy for different users. Furthermore, the discounting rate is unique for every user. A fact that may influence the useful life is the duration of use per day and the number of switches. Costs for the replacement of the lamps, potential inventory costs, and overheads (e.g., for ordering lamps and acquiring information) are difficult to calculate; they are excluded from this case study. One also has to deal with the different useful lives of the objects analyzed (i.e., a benchmark has to be defined). The company supplying the data for the case study intended to show the advantages (economic and ecological) of its new products to potential customers.

The results of the case study (see, for example, Figure 7.5) were included in the advertisement of the company.

In addition, the absolute volume of the LCC costs of energy-saving lamps is low compared to that of the incandescent lamp (see also Okada et al. 2002). For the energy-saving lamp type 1, the LCC costs are approximately 32% of those of an

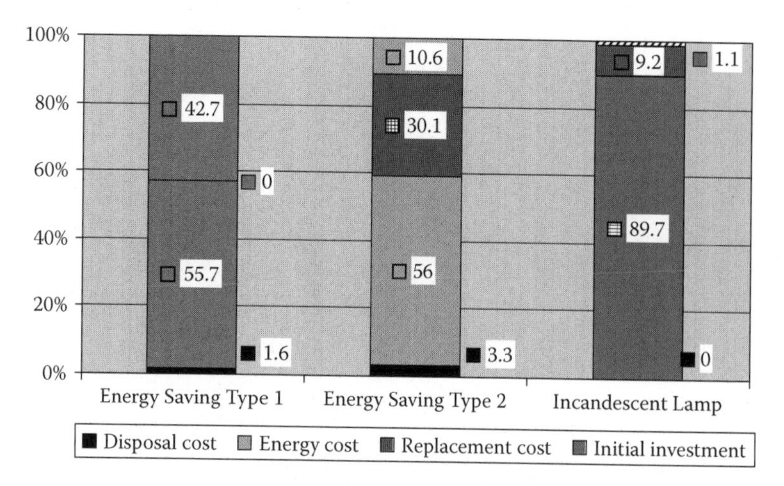


**FIGURE 7.5** Structure of life cycle costs for different lamps.

**TABLE 7.7**
**Breakeven calculation for the type 1 (energy-saving) versus type 2 (incandescent) lamp**

| Parameter | Breakeven point |
|---|---|
| Price | 91 615 € |
| Energy consumption | 71.6 W |
| Disposal cost | 83.30 € |
| Monthly interest rate | 19% |
| Energy price | 0.0075 €/kWh |

incandescent lamp (1808.68 €), and for the energy-saving type 2 lamp, they are 64% (3595.06 €) of the LCC costs (5614.51 €).

The data can be further assessed using sensitivity and/or breakeven analysis. Table 7.7 shows the results of such a breakeven analysis. Here the question is analyzed when an energy-saving lamp (type 1) is no longer better than an incandescent lamp.

The LCC reveals that there may be a financing problem in the short run (i.e., the acquisition costs are higher), and this could influence liquidity. Furthermore, it was shown that LCC is a useful instrument to demonstrate economic consequences in the long run and that it might be combined with an ecological assessment using the same data.

### 7.4.5 Overview of the Tools Used

The conventional life cycle costing approach has been applied, as summarized in Chapter 1. The results are calculated using Excel spreadsheets. The results are further verified using breakeven and sensitivity analyses.

## 7.5 DOUBLE-DECK CARRIAGE FLOOR (BAHNKREIS PROJECT)

### 7.5.1 Summary

The Bahnkreis project (Fleischer et al. 2000) was concerned with the development of a method to operate railways in a sustainable way through the use of internal life cycle cost and environmental assessments. The study involved the gathering of interested parties and stakeholders through the life cycle of railway vehicles such as railway consultants and scientists, railway-operating companies, as well as railway-producing companies. Specifically, this environmental LCC investigated life cycle costs plus environmental impacts, via a life cycle assessment, of a double-deck carriage floor from a specific train system operating in Germany.

### 7.5.2 Definition of the Case Study

The investigation was motivated by the fact that decisions needed to be made on the construction of the floor and on cleaning, maintenance, and disposal of the carriage. It involved personnel from the railway carriage–producing company and the operating company.

The floor in a double-deck railway carriage (i.e., load-bearing frame, cover, finish, plywood, and aluminum structure) was investigated. Figure 7.6 provides an illustration of the railway carriage. The floor was constructed from plywood with an aluminum sandwich profile. The functional unit was 1 floor of a specific train operating in the Ruhrgebiet-Aachen area in Germany, with an annual operating distance of 377 238 km, and operating for 30 years. The floor measures approximately 42.5 $m^2$ and comprises a rubber coverage on a weight-bearing construction. A life cycle inventory and life cycle costing were performed in parallel with the total life cycle costs, arriving at 123 374 € when discounted by 5%, and 248 000 € for the nondiscounted costs (as is the norm for environmental LCC). The costs considered were production, operation, cleaning, maintenance, modernization, and disposal. The purchase cost of materials was found to be 3% of the overall life cycle costs, while cleaning and maintenance costs over the life cycle were 75% and use costs (allocated energy consumption due to the weight of the floor) contributed 16%. Other information collected was the reliability of floor covers to determine maintenance frequency.

The approach taken was to assess the life cycle costs on the basis of a life cycle inventory. Therefore, the similarity in the functional unit and system boundary to LCA, as well as the supplemental environmental analysis, renders this case an environmental LCC. The materials within the inventory were multiplied with specific prices, including working and machine hours in the inventory. Specific prices per person-hour and machine-hour (distinguished by type of machine and type of work) were also included. All other costs were allocated on the level of processes in the inventory. The time (years) for each process was estimated. To do so, starting from a maintenance regime (maintenance processes at scheduled time or distance intervals), with stochastic additions by unplanned repairs due to component failures, and completed by duration defined for every process, the inventory was modeled over time. Inventory costs were aggregated per year and then discounted (5% rate) per

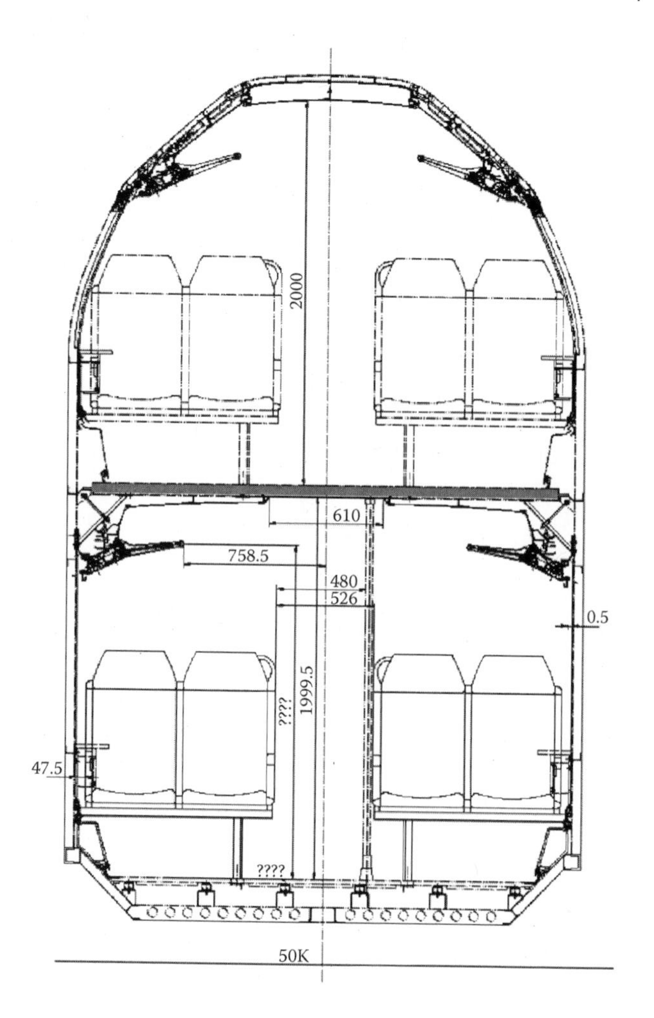


**FIGURE 7.6** Floor in a double-deck carriage operating in Germany. *Source*: Picture courtesy Bombardier Transportation.

year. A software program was developed to enable the calculations. Figure 7.7 shows combined results for the climate change indicator results and life cycle cost figures for the floor, with a lifetime of 30 years.

The study was conducted by internal company sources in conjunction with an external consultant and a university in Germany during 1998–2000. Personnel who were involved included the railway consultants and scientists, the railway-operating company, and railway-producing companies.

### 7.5.3 Entry Gate and Drivers

The project's entry gates comprised senior management and senior construction engineers who were supported by external consultants and by a public project sponsor.

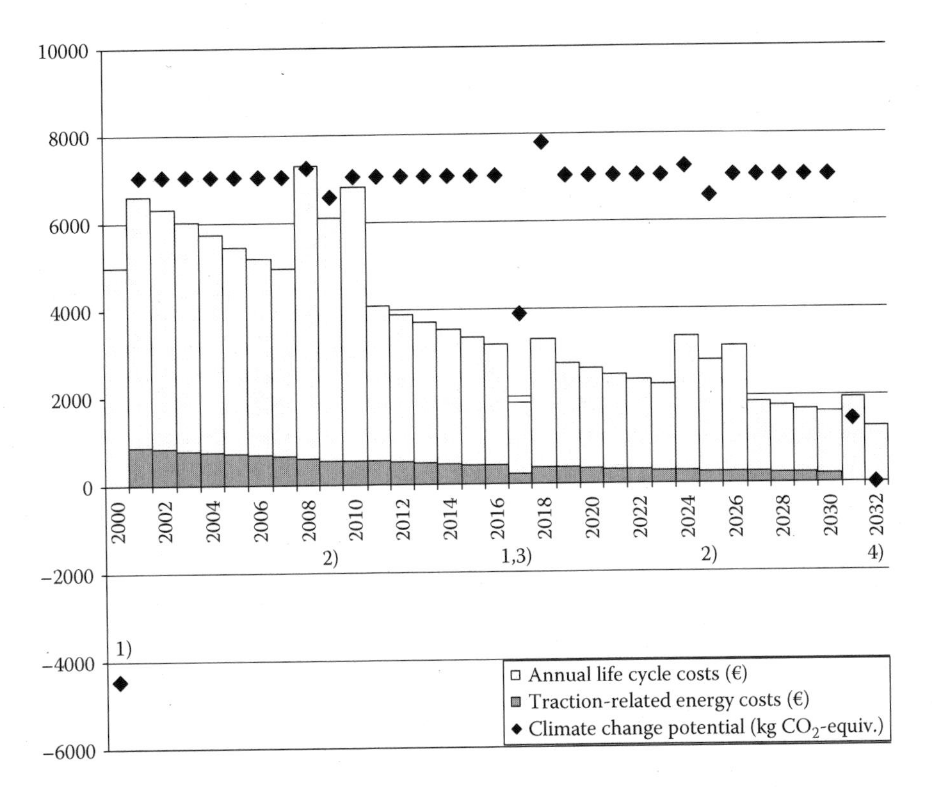


**FIGURE 7.7** Results of life cycle costs (€) and climate change potential per year, for the wooden floor variant. *Note*: Costs are discounted by 5%. 1) Negative potential due to incorporated $CO_2$, 2) revision of the train, 3) modernization and reproduction of the floor, and 4) disposal (waste incineration plant).

In the railway sector, purchase costs make up only a small portion of the overall costs of ownership and of the life cycle costs. Hence when answering a call for tender, providing and guaranteeing life cycle costs in addition to purchase prices is becoming increasingly common, and this can be viewed as the driver for the case. A reason for launching the project was a fragmentation of individual solutions in industry and a need for networking between industry, consulting, and railway operators. The environmental assessment was added due to a general interest in the industry and was also motivated by the project sponsor. In the case study, a lightweight metal frame was clearly preferable to a traditional wood construction, from both economic and environmental aspects.

### 7.5.4 Implementation

Several barriers existed with implementation. First, cost data are sensitive data and their exchange along the supply chain can be a problem. Different cost definitions and allocations of costs can hamper consistent decision support, as can the lack of adequate tools for providing accepted and sound decision support figures. The process to achieve change in the project included intensive, and open, communication between academia, those involved in methodology, consultants, construction engineers, and middle management. It included the development of a tool for calculating the life cycle of a train, over time, consistently for LCC and LCA. By the end of the project, understanding between stakeholders was obtained. Continuity within the project team was an issue, which hampered communication progress. A tool was developed and used by the project partners. The tool incorporated methods for a consistent coupling of LCC and LCA. The applicability of the tool and the methods developed could be demonstrated with the case study, which showed a clear preference for a new construction variant.

A combination of different assessment methods allows coping with different backgrounds and interests in interdisciplinary projects and allows answering purchase or product design questions that influence a multitude of different areas such as costs and the environment. It was important that there existed a general understanding and trustworthiness in the tool's result prior to using it to make decisions.

### 7.5.5 Overview of the Tools Used

The tools applied in this case study were life cycle assessment according to ISO 14040/44 (2006), Siemens' compass method, life cycle costing, and relative costing according to VDI (1984). For more information on the software tools developed, refer to Ciroth (2002) and Ciroth et al. (2003). The later includes specifications of the Siemens' compass method as used in the Bahnkreis project.

## 7.6 WASHING MACHINES

The case study described in this section is based on real data, as were the preceding 4 applications. This case serves as a basis for the idealized case study boxes that can be found throughout the book. In order to prepare boxes that illustrate the wide range of LCC applications and the impacts of different methodological choices, the case at hand was partly extended with assumptions of the authors (Table 7.8). The real case documents a conventional LCC, while the hypothetical extensions lead to environmental and societal LCC, respectively.

The following description refers only to the underlying real case on a summary level. A detailed description of the complete study can be found in Rüdenauer et al. (2004) and Rüdenauer and Gensch (2005a). A description of the study that was another basis for the case study boxes can be found in Rüdenauer and Grießhammer (2004).

### 7.6.1 Summary

The study "Eco-Efficiency Analysis of Washing Machines — Life Cycle Assessment and Determination of Optimal Life Span," conducted by Öko-Institute for the manufacturers Electrolux-AEG Hausgeräte GmbH and BSH Bosch und Siemens Hausgeräte GmbH (Rüdenauer et al. 2004), aimed to answer, among others, the following questions:

1) What are the environmental impacts of a washing machine over its whole life cycle (production, distribution, use, and end-of-life treatment)?
2) Does it make sense from the economic and the environmental points of view to further use an old washing machine, or is it better to buy a new one?

**TABLE 7.8**
**Overview of all washing machine case study boxes: real case study and hypothetical extensions**

| Case study box | Section | Theme | Real case study | Hypothetical additional features of environmental and societal LCC |
|---|---|---|---|---|
| 1 | 1.4.1 | Goal and scope definition | Consumer oriented, government and society oriented (Rüdenauer and Grießhammer 2004) | Manufacturer, supplier, and end-of-life service provider–oriented |
| 2 | 2.3 | Cost categories | Main cost categories, 3rd and 4th level (Rüdenauer and Grießhammer 2004; Kunst 2003) | Some cost categories of manufacturer and end-of-life service provider |
| 3 | 2.4 | Perspectives | Consumer perspective (Rüdenauer and Grießhammer 2004) | Different perspectives of manufacturer, and government and society |
| 4 | 2.6.1.1 | Long-term discounting of results | Not available | Discounted result |
| 5 | 3.3.3 | Calculation with discounted cash flow | Discounted cash flow (Rüdenauer and Gensch 2005a) | Not necessary |
| 6 | 3.3.4 | Calculation of life cycle costs | Calculation equation not revealed | Equation for calculation and aggregation of life cycle costs |
| 7 | 3.4 | Different types of LCC | Conventional LCC (Rüdenauer and Grießhammer 2004), LCIA (Kunst 2003) | Environmental LCC, parts of conventional LCC, and societal LCC |
| 8 | 3.5.2.2 | Input–output analysis | Internal cost (Kondo and Nakamura 2004) | Carbon tax considered |
| 9 | 4.4 | Externalities and internalizing externalities | Not available | Externalities |
| 10 | 5.2 | Presentation of LCC results | Consumer perspective (Rüdenauer and Grießhammer 2004) | Whole life cycle |

To answer the first question, an existing LCA study of 1995 has been updated, and additionally an LCC has been conducted. To answer the 2nd question, the further use of washing machines of different ages has been compared with the acquisition and use of a new washing machine bought in 2004.

### 7.6.2 Definition of the Case Study

Due to technological advances during the past 10 to 15 years, in both washing machines and detergents, significant reductions of energy and water consumption could be realized in the field of private laundry. LCA has shown that the use phase is dominant compared to the production or end-of-life phase of washing machines. However, these LCA results are relatively old. In the meantime, several parameters affecting the results of the LCA may have changed: for example, machine technology has changed (more plastic and electronic components), and through the WEEE directive the end-of-life management of washing machines is modified and consumer behavior (choice of washing temperature and loading) has also evolved.

Against the aforementioned background, 2 major tasks of the study were to update the LCA for washing machines and to additionally calculate the life cycle costs (task 1) and to analyze if it makes sense to further use an old washing machine or if it is better (in environmental and economic terms) to buy a new one (task 4 of the study).* In a supplementary study (Rüdenauer and Gensch 2005a), task 4 was refined with more detailed data, and several sensitivity analyses were conducted.

For task 1 (which serves as a base model for task 4), the environmental impacts and costs were calculated for a "current" washing machine model of 2004. The functional unit was defined as "washing of 8080 kg of laundry (i.e., the amount of laundry that can be washed within the life span of a washing machine) in a private household of 3 people." The life cycle costs were calculated under private households' perspective. Therefore, the costs for production and delivery of the washing machine were not examined in detail. Instead, the transfer price of a washing machine was taken. Table 7.9 provides an overview of the considered costs.

The costs over the total life span of 2000 washing cycles were calculated (without discounting of the future costs) as the first deliverable or task in the project. In the case of a household of 3 people (175 washing cycles per annum), this results in 11.4 years.

Figures 7.8 and 7.9 show the cumulative energy demand (CED, representing 1 environmental aspect) and the life cycle costs as results of the conducted LCA and LCC. Table 7.10 summarizes the cumulative energy demand and the life cycle costs. The most obvious result from this comparison is the difference of the contribution of the production or acquisition, the energy supply, and the water supply to the total CED and the total costs respectively.

The environmental impacts and the costs of the use of existing washing machines in stock were compared (task 4 of the project) to those of the production or purchase

* Two other tasks (which are not of particular interest with regard to the examples chosen in the case study boxes) were to compare the environmental and economic consequences of using a washing machine with a rated capacity of 5 kg with those of using a larger washing machine (task 2), and to determine the optimal life span of a washing machine regarding the next approximately 20 years (taking into account potential future developments; task 3).

**TABLE 7.9**
**Overview of the costs considered in the washing machine case study**

| Cost category | Cost per unit | Comment |
|---|---|---|
| Acquisition costs | 500 €/machine | Internet research (in 2004) |
| Energy supply | 0.18 €/kWh | Own compilation (in 2/2003), linear increase to 0.249 €/kWh in 2020 |
| Water supply (including wastewater treatment) | 4 €/m³ | Own assumption, increase by 2% per annum |
| Detergent supply | — | Not considered, as being irrelevant for task 4 |
| Repair and maintenance | — | Not considered due to data uncertainty |
| Disposal | 0 € | Disposal free of charge in most German cities; according to WEEE, the costs will be borne by manufacturers. |

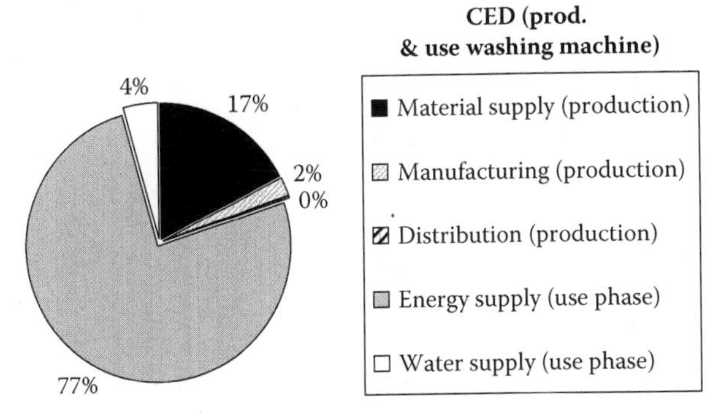


**FIGURE 7.8** CED of the life cycle phases of a washing machine (without considering recycling credits).

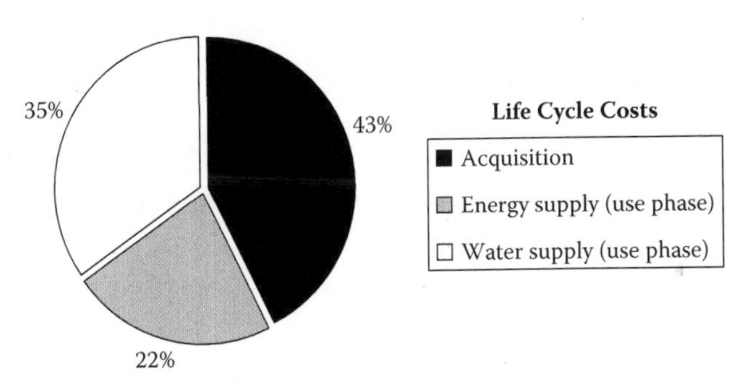


**FIGURE 7.9** Life cycle costs of a washing machine.

and the use of a "current" washing machine. Therefore, the time required to save the additional impacts and costs caused by the acquisition of a new washing machine with potentially lower environmental impact and costs during the use phase was analyzed. Here, in addition to the washing process, the drying of clothes was also included, as the energy consumption of the drying process is influenced by the spin speed of washing machines, which also increased during recent years.

**TABLE 7.10**
**CED and LCC of the production or acquisition and use of a washing machine**

| Item | CED (per washing machine) | LCC (per washing machine) |
|---|---|---|
| Material supply (production) | 3074 MJ | — |
| Manufacturing (production) | 406 MJ | — |
| Distribution (production) | 28 MJ | — |
| Subtotal production or acquisition costs | 3508 MJ | 500 € |
| Energy supply (use phase) | 13 248 MJ | 262 € |
| Water supply (use phase) | 711 MJ | 406 € |
| Subtotal use phase | 13 959 MJ | 746 € |
| Total | 17 567 MJ | 1221 € |

The functional unit was defined as "washing and drying of annually 707 kg of laundry (i.e., the annual amount of laundry in a household of 3 people) over a period of 10 years in a private household of 3 people." Five options were compared: the further use of 4 washing machines of different ages (bought in 1985, 1990, 1995, and 2000), and the acquisition and use of a new one in 2004. For each option, the environmental impacts and the life cycle costs were calculated on an annual basis (per year). These annual values are then cumulated to give the total environmental impacts and costs after 1, 2, 3, and up to 10 years of use. All future annual costs (costs for energy and water demand for the washing process, costs for the energy demand of the drying process, and, where applicable, the acquisition costs) that occur in the years between 2004 and 2025 were discounted with an annual discount rate of 5% to give the net present value (NPV) in 2004.

Table 7.11 and Figure 7.10 show the cumulated energy demand of the 5 options. It can be seen that the differences of the CED between the alternatives are quite small. Nevertheless, the additional CED for the production of a new washing machine is amortized in all cases within the following 10 years. The years where the acquisition of a new washing machine in 2004 is amortized are highlighted in gray in the tables.

Table 7.12 and Figure 7.11 summarize the life cycle costs for the aforementioned 5 options. One can observe that the life cycle costs skewed toward longer time frames than the CED owing to the fact that there is an acquisition cost that influences the economic (LCC) variable at the outset, though it does not apply to the environmental (CED) impact to such a high extent. This LCC demonstrates the need for consideration of marginal costs (i.e., given the fact that the washing machine functions and alternatives are compared).

### 7.6.3 Entry Gate and Drivers

Both studies were commissioned by the household appliances industry, and hence the entry gate was from the private sector. The first initiative started on a national

**TABLE 7.11**
**Cumulated CED for the use of an old or new washing machine**

**CED (in MJ) (base case)**

| | 2004 | 2005 | 2006 | 2007 | 2008 | 2009 | 2010 | 2011 | 2012 | 2013 |
|---|---|---|---|---|---|---|---|---|---|---|
| **1985** | 6.937 | 13.860 | 20.766 | 27.658 | 34.534 | 41.395 | 48.240 | 55.070 | 61.884 | 68.684 |
| **1990** | 6.436 | 12.857 | 19.264 | 25.657 | 32.036 | 38.400 | 44.750 | 51.086 | 57.408 | 63.715 |
| **1995** | 6.161 | 12.308 | 18.441 | 24.561 | 30.667 | 36.759 | 42.838 | 48.903 | 54.954 | 60.992 |
| **2000** | 5.675 | 11.338 | 16.988 | 22.625 | 28.250 | 33.862 | 39.462 | 45.049 | 50.624 | 56.186 |
| **2004** | 7.994 | 13.260 | 18.516 | 23.759 | 28.991 | 34.211 | 39.419 | 44.616 | 49.801 | 54.975 |

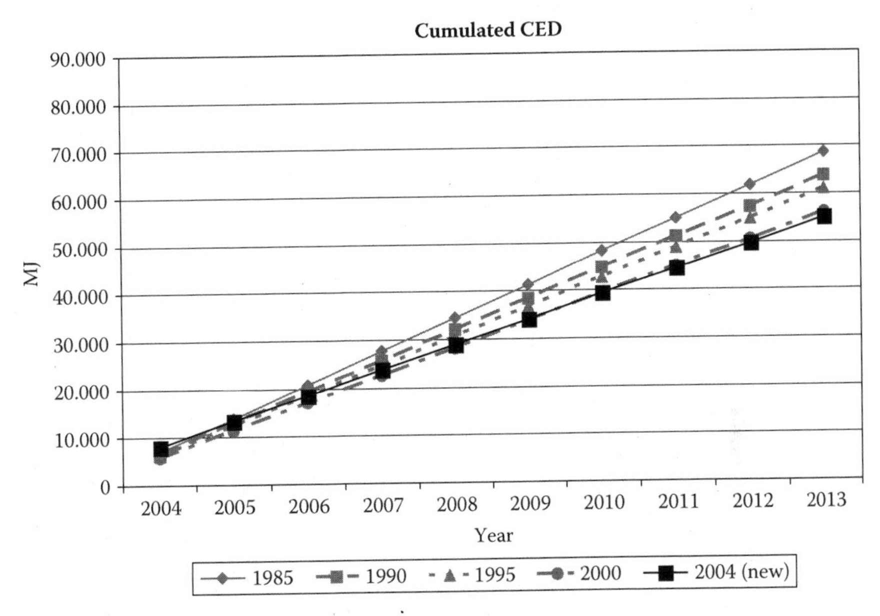


**FIGURE 7.10** Cumulated primary energy demand: old versus new washing machine. *Note*: The depicted alternatives are "1985," "1990," "1995," and "2000": further use of the existing washing machine manufactured (and purchased) in the respective year; and "2004": purchase of a new washing machine in 2004.

(German) level, then the topic was extended to the European level. Tasks 1 and 4 were meant to evaluate the benefits of an accelerated replacement of large household appliances in stock under environmental and economic perspectives. The authors believe the results could foster an early replacement initiative by industry or government (e.g., support of the replacement of the old appliance by publicly funded rebate schemes) on a European level. The driver, therefore, for the project was the industry sector with a view toward a stakeholder approach to improving the environmental portfolio of the product. The drivers also include a desire for public communi-

**TABLE 7.12**
**Cumulated life cycle costs for the use of an old or new washing machine (with discounting)**

**Costs (in Euro) (base case, with discounting)**

| | 2004 | 2005 | 2006 | 2007 | 2008 | 2009 | 2010 | 2011 | 2012 | 2013 |
|---|---|---|---|---|---|---|---|---|---|---|
| **1985** | 210 | 410 | 600 | 790 | 980 | 1160 | 1330 | 1500 | 1660 | 1820 |
| **1990** | 180 | 350 | 520 | 680 | 840 | 1000 | 1140 | 1290 | 1430 | 1570 |
| **1995** | 160 | 310 | 450 | 600 | 740 | 870 | 1000 | 1130 | 1250 | 1370 |
| **2000** | 140 | 270 | 400 | 520 | 640 | 760 | 870 | 980 | 1090 | 1190 |
| **2004** | 620 | 740 | 850 | 970 | 1070 | 1180 | 1280 | 1380 | 1470 | 1570 |

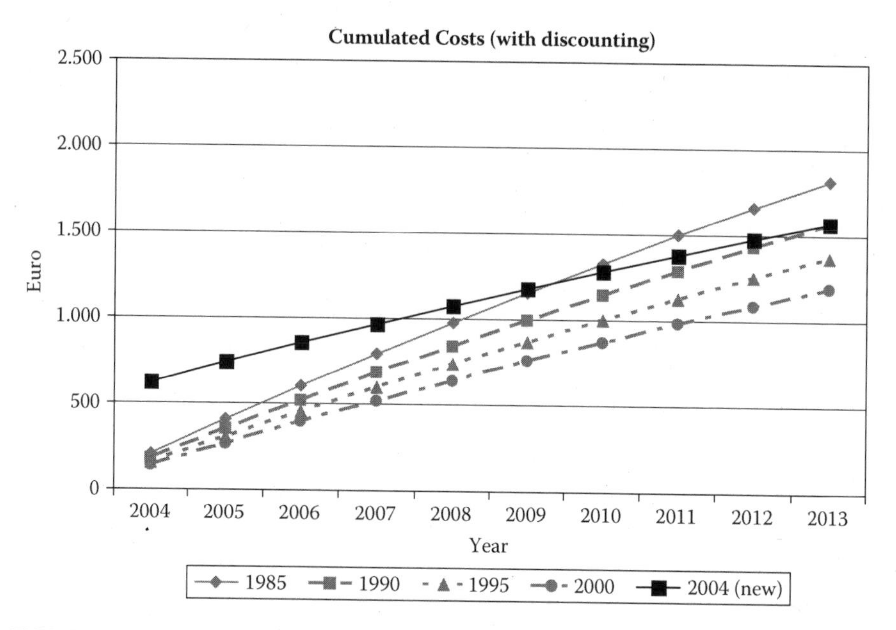


**FIGURE 7.11** Cumulated costs: old versus new washing machine (with discounting). *Note*: The depicted alternatives are "1985," "1990," "1995," and "2000": further use of the existing washing machine manufactured (and purchased) in the respective year; and "2004": purchase of a new washing machine in 2004. All future costs are discounted with a discount rate of 5% to give the net present value (NPV) in 2004 (see also explanation above).

cation. The motivation for this study and more environmentally friendly washing machine services can, therefore, be viewed as a balance between technology push and informed opinion-based demand.

### 7.6.4 Implementation

In addition to the washing machine case, cold appliances were also investigated in a subsequent study (Rüdenauer and Gensch 2005b). With the results of all 3 studies, the household appliances industry started to inform public authorities about the implications of an accelerated replacement of large household appliances.

### 7.6.5 Overview of Tools Used

The environmental impacts were assessed via the cumulated energy demand, and a conventional LCC approach was used to calculate the costs from the perspective of the consumer. Cash flows within the lifetime of the washing machine were discounted.

## 7.7 HYPOTHETICAL CASE: A HIGH-CAPACITY GLASS CABLE NETWORK FOR DATA TRANSMISSION

### 7.7.1 Summary

The application of societal LCC, with variants in the dimensions, will be demonstrated with a hypothetical example: the costs of a door-to-door high-capacity glass cable network. The example discusses the LCC from the perspective of different actors, including the producer, service provider, consumer, and government.

### 7.7.2 Definition of the Case Study

The goal and scope of the study may be the choice for this cable system or, for example, a lower capacity system combined with copper wiring and ADSL type of data transmission, or high-frequency radio transmission systems. Each option has its specific time profile. This example refers to 1 option only, the high-capacity glass cable network. This example has been chosen since it can be used to demonstrate the differences between the various options in LCC. The costs of the cable system are shown in Table 7.13 (the reader is cautioned to note that the data are hypothetical, as are the environmental costs as indicated in monetary terms).

Table 7.13 distinguishes 5 cost categories: construction, maintenance, and operation; payments to investors; user fees; government subsidies and VAT; and environmental damage. These categories do not relate to 1 specific level of cost categories; actually, only the construction, maintenance, and operation costs (and, in the social cost view, also the environmental costs) are real costs as involving the use of scarce alternatively applicable resources. All other costs here are transfer payments between different cost bearers.

The costs are calculated for 5 time periods: a 5-year building period, 2 operational periods (respectively, 25 and 19 years), 1 year in which the system is closed for major overhaul (not so realistic but to add possible complexity to the computation), and 2 years of demolition. Table 7.13 displays the costs per year in the indicated period. Negative numbers indicate that for the entity-bearing cost, there are net proceeds. A government subsidizing a high-capacity data transmission system, for example, may well have tax proceeds from its exploitation that are higher than this subsidy. Please note that of the 52 years where costs accrue, the system is functioning for only 44 years.

**TABLE 7.13**
**Life cycle costs of a high-capacity glass cable network[a]**

| Years | Construction, maintenance, and operation[b] (A) | Payment to investors (B) | User fees (C) | Government subsidy or VAT (D) | Environmental damage (E) |
|---|---|---|---|---|---|
| 1 to 5 (building) | 500[c] | –350 | 0 | 150 | 150 |
| 6 to 30 (operational) | 25 | 130 | 200 | –32 | 5 |
| 31 (closed for maintenance) | 200 | 130 | 0 | 0 | 20 |
| 32 to 50 (operational) | 30 | 130 | 200 | –32 | 5 |
| 51 to 52 (demolition) | 100 | 0 | 0 | 100 | 20 |
| Totals (no discounting) | 4095 | 4100 | 8800 | –458 | 1030 |

[a] The case uses a constant inflation rate of 3% per annum, capital rent of 7%, a social time preference of 3%, a VAT of 20%, and a functional life of 44 years. Data on the inflation, capital rent, and social time preference are needed to compute the discount rate for market, alternative, and social approaches. The VAT rate is needed to calculate the VAT on the user fee.

[b] The operation of the system is included; operation of connected user–consumer hardware is not included.

[c] All monetary values are in millions of euros per year.

### 7.7.3 Entry Gates and Drivers

Table 7.13 can be used to calculate the costs (or negative costs, which are revenues) of the system for various persons or groups. Table 7.14 thus simulates various perspectives and shows the net costs for some groups of cost bearers.* The producer is only responsible for the construction of the hardware, especially cables, which will cost 2500 million euros, or MEuro (5 × 500 MEuro). The users of the system will pay an annual fee of 200 MEuro. They only have to pay when the system is operational, that is, for 44 years (44 × 200 MEuro = 8800 MEuro). There will not be an elaborate discussion on the net costs of these 2 cost bearers as they are the most simple ones owing to the fact that they do not receive any money.

The calculation of the net cost of the service provider is more disputable. The costs of the service provider consist of the building, maintenance, and operation costs; the payments to the investors; and taxes to the government (VAT). The user's fees are his or her benefits (negative costs). Here, it can be discussed if the VAT payments and user fee benefits should be included in the net costs. The same holds for the

* In a hypothetical case, there are clearly neither entry gates nor drivers, and, therefore, the various actors' perspectives have been evaluated.

**TABLE 7.14**
**Overview of net costs for some (groups of) cost bearers (no discounting)**

| Years | Producer | Users | Service provider | Users and service provider | Users, service provider, and investors | Government | Society, no environmental costs | Society, including environmental costs |
|---|---|---|---|---|---|---|---|---|
| | A | C | A + B – C – D | A + B – D | A – D | D | A | A + E |
| 1–5 | 2500 | 0 | 0 | 0 | 1750 | 750 | 2500 | 3250 |
| 6–30 | 0 | 5000 | –325 | 4675 | 1425 | –800 | 625 | 750 |
| 31 | 0 | 0 | 330 | 330 | 200 | 0 | 200 | 220 |
| 32–50 | 0 | 3800 | –152 | 3648 | 1178 | –608 | 570 | 665 |
| 51–52 | 0 | 0 | 0 | 0 | 0 | 200 | 200 | 240 |
| Totals | 2500 | 8800 | –147 | 8653 | 4553 | –458 | 4095 | 5125 |

*Note:* The letters in this table (A to E) show which of the cost categories from Table 7.13 are used to compute the costs. Please note that this is 1 of the many possible tables that can be made based on the definition of net costs.

groups of cost bearers containing the service provider: the users and service provider; and the users, service provider, and investors. When looking at the users and service provider together, fee payments can be left out of the account as in-group transfer payments. What remains are building and maintenance costs, payments to the investors, and VAT payments. Whether this tax payment is a relevant money flow is a matter of taste, as it may well be assumed that alternative activities would generate taxes as well. It surely makes public investment projects look nicer. When including the investors and looking at the users, service provider, and investors, the payments to and from the investors are left out of the account. What then remains are only the maintenance and operation costs, as well as the VAT payments (again, disputable).

The government pays a part of the building costs (the rest is paid by the investors through the service provider) and all demolition costs, and receives VAT on the fee payments. Again, the inclusion of the tax payments can be discussed. When looking at society, all transfer payment cancel out and hence are to be neglected. When excluding the environmental costs, only the construction, maintenance, and demolition costs are used. Naturally, the environmental costs are included when estimating the social costs.

Table 7.14 reflects that the principal life cycle cost components arise in the use phase, with the production phase also quite significant. The cost of the service provider, while negative, is negligible; and the government subsidy is, while above the threshold for costing, a relatively minor contribution. The societal costs, either with or without environmental costs internalized, are approximately half of the use phase monetary burden. Hence, one can look at the overall environmental overhead of a data network as being approximately 50%.

### 7.7.4 Implementation

In order to illustrate the implementation phase, the perspective of the producer is arbitrarily selected. Table 7.15 shows, as an example, the nominal and discounted costs for the producer, which diverge substantially in the relatively short period of 5 years.

**TABLE 7.15**
**Nominal (= 0%) and discounted costs for the system builder**

| | Discounting | | |
|---|---|---|---|
| **Year** | **0%** | **3%** | **7%** |
| 1 | 500 | 485 | 467 |
| 2 | 500 | 471 | 436 |
| 3 | 500 | 457 | 408 |
| 4 | 500 | 444 | 381 |
| 5 | 500 | 431 | 356 |
| Total | 2500 | 2289 | 2050 |

### 7.7.5 Overview of the Tools Used

As the case is hypothetical, it should be pointed to the tools that would, at this stage, most likely be of service in carrying out a societal LCC. These include the net present value for the estimation and comparison of costs, as well as average yearly costs for nondiscounted options. Importantly, one should note that for societal LCC, it is not so much the tools, which are traditional financial instruments, which are difficult, but rather the assurance that the complete set of data has been collected, with overlap and double counting avoided. Therefore, diligence in identifying subsidies and long-term environmental costs is essential. A good survey of the LCC tools, with many problems treated very similarly to those in LCA inventory modeling, is given by Dhillon (1989). Other texts describing in detail the LCC toolbox are Badgett et al. (1995), USEPA (1996), and Gray and Bebbington (2001).

## 7.8 PASSENGER CAR

### 7.8.1 Summary

Price is an important criterion in private consumers' purchasing decisions. However, often the subsequent costs for operating the product exceed the initial purchasing costs and strongly depend on certain product characteristics. The comparison of the life cycle costs of products, therefore, usually provides a more realistic comparison of alternatives. To assist consumers in placing sustainability into purchasing decisions, the EcoTopTen campaign (http://www.ecotopten.de; EcoTopTen 2007) provides information on the environmental impacts and the life cycle costs from a private household's perspective of a variety of product groups, including all passenger cars. These are categorized in the best category of the "Auto-Umweltliste," a ranking established by a German association for sustainable mobility (Verkehrsclub Deutschland, or VCD; www.ecotopten.de; Gensch and Grießhammer 2004; Öko-Institut 2006). In the present conventional LCC, the life cycle costs of 3 cars of the smallest size category defined (Öko-Institut 2006) are compared. The selection of the 3 cars is due to their purchasing price and utility, which were virtually identical.

### 7.8.2 Definition of the Case Study

The LCC analysis of the passenger cars was conducted from the perspective of a private, German household. The analysis was made for a holding period of 4 years, including the purchase and the use of a new car and its resale after that period. The functional unit was defined as "the supply of an annual mileage of 12000 km by a passenger car." Data were collected with a time-related coverage of 2005 for Germany as a geographical scope. The calculation procedure is comparable to that of the "ADAC Autokostenrechnung," a conventional LCC tool developed by Allgemeiner Deutscher Automobil (ADAC), the German automobile club (ADAC 2005). The assumptions are defined by Öko-Institut. The LCC described here compares the costs of the cars "Opel Corsa 1.0 Twinport Ecotec," "Fiat Punto 1.2 8V," and "Citroën C2 1.1."

As the main cost categories, the acquisition expense (depreciation of car, imputed interest of purchase price, and initial costs for transfer), fixed costs during the use phase (tax and insurance, rent for garage, parking fees, etc.), operating costs (for fuel, costs, washing, and general care), and maintenance (e.g., tire wear and inspection) were included. In the following, the calculation of these 4 cost categories is described in more detail.

### Acquisition Costs

Depreciation (straight line) was calculated from the purchase price according to the recommended retail price of the manufacturer and the resale price for the used car after 4 years and an annual mileage of 12 000 km according to the Deutsche Automobil Treuhand.

With reference to the purchase price, an imputed interest of 5% is assumed. Finally initial costs for transfer, registration, and the number plate are taken as 500 €. Table 7.16 summarizes the assumptions and the resulting costs.

### Fixed Costs

Car-specific tax and insurance costs are considered as fixed costs during the use phase. Additionally, further nonspecific fixed costs are included in the calculations, such as rent for a parking garage, parking fees, maps, costs for (compulsory) general inspection, and so on. Table 7.17 provides an overview of the resulting costs.

### Operating Costs

Operating costs, shown in Table 7.18, comprise the costs for fuel, calculated from annual consumption and fuel prices; costs for oil; and costs for washing and general

**TABLE 7.16**
**Assumptions and annual costs with regard to the acquisition**

| | Opel Corsa 1.0 Twinport | Fiat Punto 1.2 8V | Citroën C2 1.1Advance |
|---|---|---|---|
| **Assumptions** | | | |
| Purchase price (no extras, no discount) | 10945 € | 10890 € | 10990 € |
| Resale value | 5144 € | 4247 € | 5385 € |
| Holding period | 4 years | 4 years | 4 years |
| Annual mileage | 12 000 km | 12 000 km | 12 000 km |
| Imputed interest | 5% | 5% | 5% |
| **Costs** | | | |
| Depreciation (per annum) | 1450 € | 1661 € | 1401 € |
| Interests (per annum) | 402 € | 378 € | 409 € |
| Transfer (per annum) | 100 € | 100 € | 100 € |
| Registration and the number plate | 25 € | 25 € | 25 € |

**TABLE 7.17**
**Annual fixed costs during the use phase**

| Costs per annum | Opel Corsa 1.0 Twinport | Fiat Punto 1.2 8V | Citroën C2 1.1Advance |
|---|---|---|---|
| Tax | 68 € | 88 € | 81 € |
| 3rd-party liability insurance | 794 € | 871 € | 871 € |
| Comprehensive cover | 691 € | 752 € | 375 € |
| Further fixed costs | 200 € | 200 € | 200 € |

**TABLE 7.18**
**Assumptions and data with regard to operating costs**

| | Opel Corsa 1.0 Twinport | Fiat Punto 1.2 8V | Citroën C2 1.1Advance |
|---|---|---|---|
| **Assumptions** | | | |
| Mileage (per annum) | 12 000 km | 12 000 km | 12 000 km |
| Fuel consumption (liter/100 km) | 5.3 | 5.7 | 5.9 |
| **Costs per annum** | | | |
| Fuel | 782 € | 841 € | 871 € |
| Oil | 7 € | 3 € | 7 € |
| General care | 120 € | 120 € | 120 € |

care. All 3 cars need “super fuel” for which a price of 1.23 € per liter is assumed (no increase assumed during the holding period of 4 years; increases may be modeled in a sensitivity analysis).

## Maintenance Costs

For repairs, costs for material and working time for repairs, tire wear, and inspections are included (Table 7.19). For all repair facilities, the same hourly rate of 60 € is assumed. The number of working hours has been set by the above-mentioned software of ADAC.

**TABLE 7.19**
**Assumptions and data with regard to maintenance costs**

| Costs per annum | Opel Corsa 1.0 Twinport | Fiat Punto 1.2 8V | Citroën C2 1.1Advance |
|---|---|---|---|
| Tire wear | 44 € | 76 € | 67 € |
| Repairs (material and labor) | 226 € | 251 € | 150 € |
| Inspection (material and labor) | 82 € | 163 € | 101 € |

## Life Cycle Costs

Figure 7.12 and Table 7.20 show the annual life cycle costs of the regarded passenger cars with an assumed mileage of 12 000 km per year and a holding period of 4 years.

When using purchase price as the sole decision metric, there is nearly no difference between the 3 automobiles. The Fiat Punto is slightly less expensive than the Opel Corsa and the Citroën Advance. However, when the life cycle costs are regarded, the Opel and Citroën are respectively 540 € and 750 € per year less expensive than the Fiat. Interestingly, the annual life cycle costs add up to some 50% of the initial purchase price, which implies that within the assumed holding period of 4 years, consumers spend roughly twice the purchase price for the acquisition and use of the car, even though they resell the car after this period for 40–50% of the purchase price (see "Resale value" in Table 7.16).

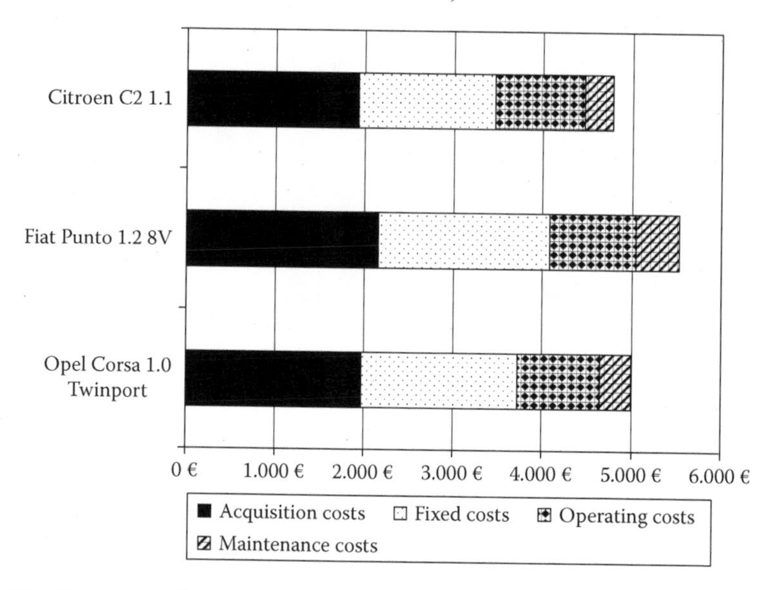


**FIGURE 7.12** Annual life cycle costs of 3 passenger cars.

**TABLE 7.20**
**Annual life cycle costs of the regarded passenger cars**

| | Opel Corsa 1.0 Twinport | Fiat Punto 1.2 8V | Citroën C2 1.1Advance |
|---|---|---|---|
| Purchase price | 10945 € | 10890 € | 10990 € |
| **Life cycle costs (per annum)** | | | |
| Acquisition costs | 1977 € | 2164 € | 1936 € |
| Fixed costs | 1753 € | 1911 € | 1527 € |
| Operating costs | 909 € | 964 € | 998 € |
| Maintenance costs | 352 € | 490 € | 318 € |
| Total annual life cycle costs | 4991 € | 5529 € | 4779 € |

Analyzing the life cycle costs, the main cost driver in all cases is the acquisition, comprising some 40% of the total life cycle costs. The fixed costs also play an important role (approximately one-third of the costs). The operating costs make up for some 20% of the life cycle costs and the maintenance costs for only 7 to 9%.

### Environmental Impacts

Figure 7.13 shows the environmental impacts of the 3 passenger cars, assessed by VCD points (an aggregate environmental score where a higher value has a lower impact; VCD stands for Verkehrsclub Deutschland, a German association for sustainable mobility) as described in Section 7.7.5. There are environmental differences between the cars, and after comparing them with the LCC results (Figure 7.12), it can be concluded that the Opel (medium life cycle costs) is preferable, environmentally, to the Fiat (high life cycle costs). The operating costs only add up to some 20% of the LCC (Figure 7.12); therefore, a lower fuel consumption would lower the LCC only to a small extent.

## 7.8.3 Entry Gate and Drivers

The initial research on sustainable products was conducted by Öko-Institut and sponsored by the German Federal Ministry of Education and Research between 2000 and 2004. The idea was based on a variety of projects carried out by Öko-Institut in the field of sustainable products and material flows, assessing the environmental, economic, and increasingly also the social performance of products and systems. The goal was to identify the most important product fields for environmental improvement in private households, to define "sustainable products" in these fields, and to communicate these products (which are mostly environmentally and economically beneficial for the consumers) via a consumer information campaign.

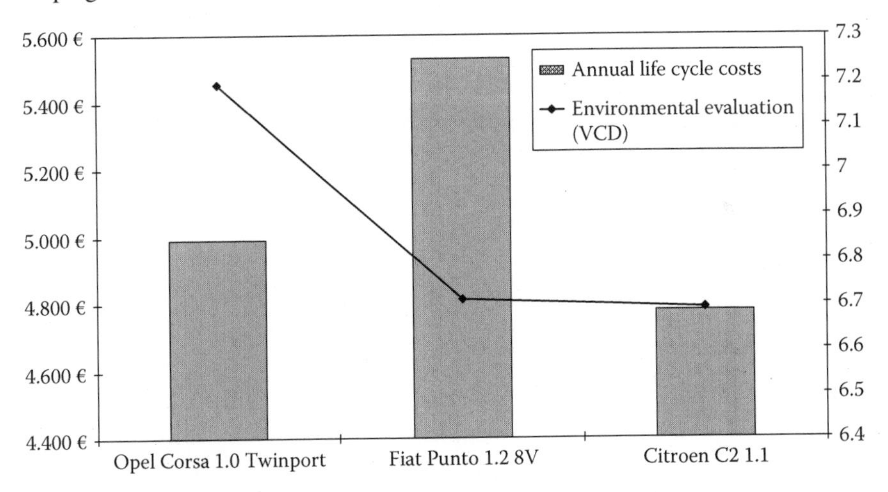


**FIGURE 7.13** Environmental impacts of 3 passenger cars (a higher value has a lower impact).

### 7.8.4 Implementation

In January 2004, the EcoTopTen consumer information campaign was launched. It is funded by the German Federal Ministry of Food Agriculture and Consumer Protection (BMELV) and the Stiftung Zukunftserbe. The campaign is based on further research funded by the German Federal Ministry of Education and Research (BMBF).

### 7.8.5 Overview of Tools Used

The environmental impacts were assessed by the VCD. The assessment is based on 3 criteria: the fuel consumption (measured by $CO_2$ emissions per km), noise emissions, and amount and type of pollutants emitted. A certain number of points has to be achieved to be scored in the environmentally best category, which is mandatory to be included in the EcoTopTen product lists.

The calculation of the life cycle costs from a private consumer's perspective is based on a conventional LCC approach. Most calculations were accomplished in Excel spreadsheets. Partly the calculations were done by a software tool developed by ADAC.

## 7.9 LIFE CYCLE COSTS OF REAL ESTATE

### 7.9.1 Summary

The challenge of life cycle costing, for real estate, is in the long life span of a building. The prognosis of costs for operation and maintenance over periods ranging from 30 to 100 years implies that scenarios must be evaluated that include maintenance and modernization. The contribution of specific cost drivers to LCC depends on the discount rate implemented as well as on the life span regarded.

### 7.9.2 Definition of the Case Study

In real estate economics, LCC is estimated to be of high relevance in decisions, with more than 60% of respondents to a 2004 survey of German real estate professionals indicating a high or an essential importance (Pelzeter 2006, 132). On the other hand, LCC was calculated for only 5% of the decisions.

This case sought to test the hypothesis that 20% of a building's life cycle costs are attributed, and realized, during design and construction. Therefore, 2 existing buildings have been analyzed in a calculation model developed by the author. The model calculates conventional LCC, neglecting any external costs.

The 1st object of the evaluation is a residential building in Berlin (Germany), Bergstraße 67, built in 2000. It is a 4-story building with an overall floor space of 410 m². The 2nd object is larger, combining shops in the basement, 3 office floors, 3 residential floors, and underground parking. This building is situated at Rheinstraße 16, Berlin (Germany), with 2500 m² rentable area, and was built in 2000 as well. The calculations were performed in 2003. At that time, costs for construction and site were known, and the costs of the first regular year of use were available.

The data in Table 7.21 show that the proportion of 20% to 80% for initial to consequential costs is only true if the time value of money is neglected in the calculation. In the case of a static calculation, the duration of the regarded life span is the main factor influencing the LCC. However, as discounting is established in the purchase and operation of a building, a dynamic calculation is necessary (Figures 7.14 and 7.15 for both buildings being investigated). The share of initial costs in a dynamic LCC reaches 50% to 60% of all costs, calculated over 90 years with

**TABLE 7.21**
**Key parameters for the 2 objects in the case study**

| Object | Bergstraße 67 | Rheinstraße 16 |
|---|---|---|
| Rentable area | 410 m² | 2500 m² |
| Costs of site minus residual value at end of life cycle | 134471 € | 1465993 € |
| Costs of construction | 459000 € | 3348000 € |
| Costs of planning | 81000 € | 301320 € |
| Further initial costs | 54000 € | 200880 € |
| Regarded life span | 90 years | 90 years |
| Discount rate (nominal) | 5% | 5% |
| Costs of operation | 379783 € | 2505011 € |
| Costs of maintenance or modernization | 243918 € | 1196155 € |
| Interval of modernization | 30 years | 30 years |
| Costs of deconstruction | 2034 € | 19789 € |

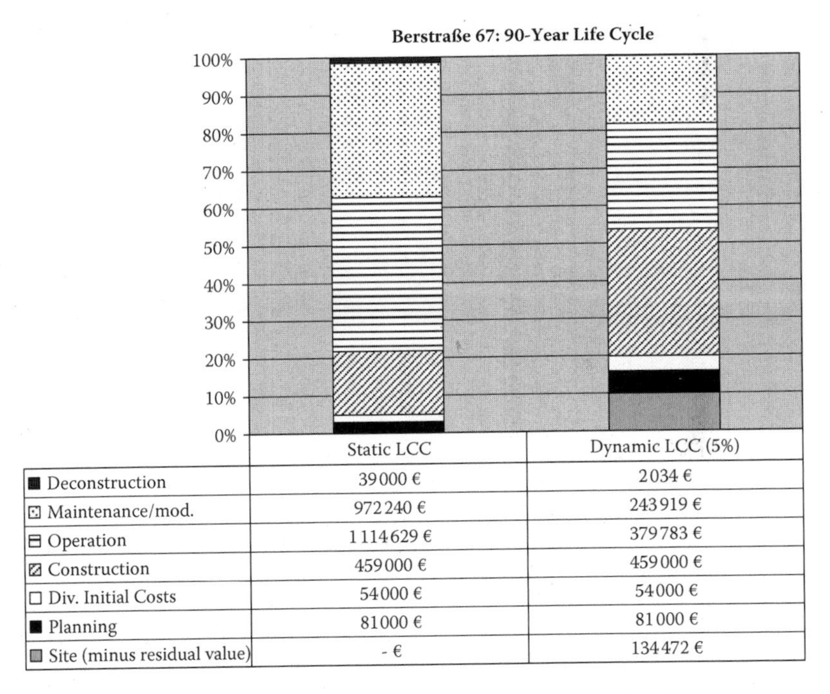


**FIGURE 7.14** Bergstraße 67: comparison of static and discounted LCC.

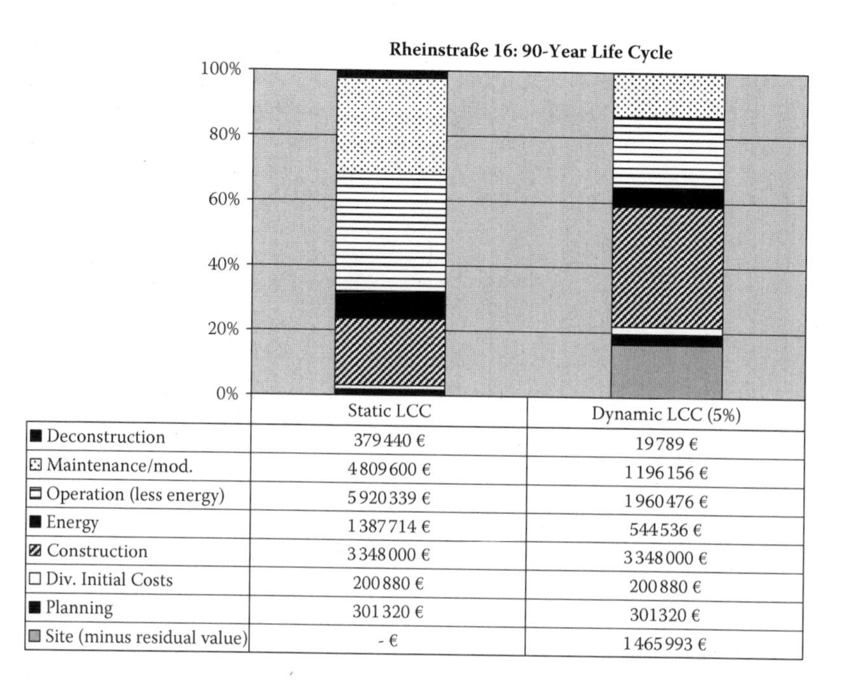


| | Static LCC | Dynamic LCC (5%) |
|---|---|---|
| ■ Deconstruction | 379440 € | 19789 € |
| ⊡ Maintenance/mod. | 4809600 € | 1196156 € |
| ⊟ Operation (less energy) | 5920339 € | 1960476 € |
| ■ Energy | 1387714 € | 544536 € |
| ▨ Construction | 3348000 € | 3348000 € |
| □ Div. Initial Costs | 200880 € | 200880 € |
| ■ Planning | 301320 € | 301320 € |
| ▩ Site (minus residual value) | - € | 1465993 € |

**FIGURE 7.15** Rheinstraße 16: comparison of static and discounted LCC.

a nominal discount rate of 5% per year (including 1.5% per year of inflation). This value applies to both residential and office buildings. A higher intensity of utilization such as is found in schools, hospitals, or industrial halls results in a smaller proportion of initial costs and a higher proportion of consequential costs. For example, a hospital may need as much as 85% of its dynamic LCC (5% discount rate, 60-year life span) for operation and maintenance or modernization.

The impression that a main part of LCC would consist in costs for energy (heating plus electricity) is also disproved by the dynamic calculation of LCC: for Rheinstraße 16, energy reaches a share of 6% in LCC, even though the specific inflation of costs for energy was assumed to be 2.5% (compared to 1.5% for other costs). It bears noting that none of the examples uses air conditioning.

There is often a debate regarding the integration of the costs of site. In a static calculation, the site is paid in the beginning and sold in the end of the life cycle at the same price; therefore, it can be left aside. In the dynamic calculation, the value of the site increases with inflation. No added value apart from inflation was assumed. However, due to discounting its present value in year 90 is much smaller than the price paid in year 1. Therefore, the difference between costs in year 1 and residual value in year 90 is defined as the costs for site. A comparison between the objects shows that the site of Rheinstraße is more expensive (close to a secondary city center of Berlin): it requires 16% of LCC versus 10% at Bergstraße.

The costs for deconstruction diminish due to discounting to a value below 1% of the LCC. Thus, the necessity of optimizing the building as a whole for a posterior recycling is not supported by economic pressure from the end of the life cycle. This is different if the less durable parts of the building are regarded (e.g., flooring materials).

### 7.9.3 Entry Gate and Drivers

The case studies have been carried out in the course of the dissertation of Andrea Pelzeter. Data for the object Bergstraße were made available by the architect and property manager Paul Ingenbleek. For Rheinstraße, generalized data derive from the owner, Commerz Grundbesitz Investmentgesellschaft (CGI), and the property manager, M+W Zander.

In order to test the influence of location, design, and environment, the buildings have been converted to a data model. Virtual variations of the existing buildings demonstrated that differences in the level of usability afford an integration of income into the calculations. These conventional LCCs in a broader sense were called "life cycle economy" (LCE), according to draft ISO/DTS 21929 (2004).

In general the real estate business is increasingly interested in calculating LCC in order to identify a positive trade-off between initial and consequential costs. Special attention to this trade-off arises in public–private partnership (PPP) or build-operate-transfer (BOT) projects, where the same investor is responsible for investment, operation, and maintenance.

### 7.9.4 Implementation

During the research, a high level of concern in relation to benchmarking LCC was noted on the part of the planners, owners, and managers of real estate (Pelzeter 2005). In 2004, the author was asked by the German Facility Management Association (GEFMA) to chair the development of a guideline for life cycle costing in facility management. The draft of guideline GEFMA 220 was published in May 2006. It recommends the implementation of standards for structuring the data on building, utilization, and derived processes, to include all costs from the point of view of the customer (tenant) and to calculate LCC as net present value, always indicating regarded life span and discount rate.

### 7.9.5 Overview of Tools Used

The calculations were run on a spreadsheet, which included a yearly prognosis of economically relevant events in maintenance, modernization, and respective consequences for the rent. All costs increased with inflation (costs for energy with specific inflation) and were discounted to year 1. The majority of respondents in the questionnaire on LCC also used customized tools in their analysis.

# 8 Conclusions

*David Hunkeler, Kerstin Lichtenvort, and Gerald Rebitzer*

## 8.1 THREE TYPES OF LIFE CYCLE COSTING

A survey, carried out at the outset of the deliberations of the SETAC-Europe Working Group on Life Cycle Costing, identified 3 types of life cycle costing. The principal differences in these methods were in their goal, scope, and system boundary, as well as in how functionality and discounting were carried out. The analysis of the 33 cases revealed that 1 type of life cycle costing represented the typical practice. "Conventional LCC" as it is termed herein, has its roots in public and military procurement and was later broadened in scope when adopted by corporations. A 2nd type of costing was strongly linked to environmental assessment and intended as a pillar of sustainability, and it is referred to as "environmental LCC." A 3rd type of costing, linked to cost–benefit analysis and encompassing social aspects, is given the nomenclature "societal LCC."

This book is focused on and hence titled *Environmental Life Cycle Costing* as this type of LCC presents a new life cycle costing methodology that is consistent with LCA and suitable as a pillar of sustainability, going beyond the well-known conventional LCC and factoring out disputable presumptions on the internalization of externalities required for societal LCC.

### 8.1.1 CONVENTIONAL LCC

Conventional LCC is, to a large extent, the historic and current practice in many governments, public organizations, as well as firms, and is based on a purely economic evaluation, considering various stages in the life cycle. Generally it is a quasi-dynamic method and includes costs associated with a product that are borne directly by a given actor, and it is usually presented from the perspective of a producer or consumer. External costs are neglected. In this sense, conventional LCC is less comprehensive in scope than systematic environmental analyses such as LCA and involves discounted costs of the results. The lowest discounting rate to be applied would be the market interest rate corrected for inflation, and the upper range is the internal rate used by organizations for their intended return on investment. This choice is left to the decision maker in analogy to managerial accounting. One should note that some external costs are already internalized (e.g., waste management in many regions of the world), while others are internalized as unforeseen costs and some may remain as external costs (e.g., caused by resource depletion, toxic releases, or social impacts). Frequently, the reference flow for conventional LCC is 1 unit of product (e.g., an automobile). The unit is convenient for calculation, though it may not be the most appropriate for a sustainability assessment.

In comparison to LCA, conventional LCC is easier to perform since data are more readily available. They are less abstract because the calculated life cycle costs are expressed in monetary units, in comparison to figures of an impact assessment. This renders the result easier to understand. In some aspects of conventional LCC application, and LCC in general, the comparison of the results of alternatives and the identification of hotspots provides the most significant input to decision making.

### 8.1.2 Environmental LCC

Environmental LCC is an assessment of all costs associated with the life cycle of a product that are directly covered by any 1 or more of the actors in the product life cycle (e.g., supplier, producer, user or consumer, or EoL actor). The complementary inclusion of externalities that are anticipated to be internalized in the decision-relevant future as well as an LCA with equivalent system boundaries and functional units are required. Environmental LCC is performed on the same basis as with LCA, with both being steady-state in nature.

Environmental LCC goes beyond conventional LCC by internalizing all real and anticipated money flows associated with the life cycle in a systematic way. It also requires a complementary LCA as well as a social assessment if the goal is to have a complete threefold sustainability assessment. Environmental LCC may use market costs (including transfer payments), alternative costs, or both. Environmental LCC is, therefore, LCA driven; thus, it uses primarily methods and concepts developed for LCA.

### 8.1.3 Societal LCC

Societal LCC uses an expanded system boundary, relative to the other types of LCC, and involves a larger set of costs, including damage costs that will, or could be, relevant in the long term. However, the monetized environmental effects of the investigated product should not be complemented by an LCA, as this procedure would lead to double counting. This quite comprehensive LCC applies a social rate of time preference type of discounting. This goes in the direction of the Brundtland Commission's requirements (1987), with approximate value below 0.1% per year, and with the precise value depending on the dominant environmental or other impacts and benefits. As was the case for conventional LCC, the method is quasi-dynamic.

In a societal LCC, externalities are included to a greater extent. Using a mixture of internal and external costing requires a precise specification of the cost bearers for which internal costing is applied, as double counting or missing costs may easily occur. Explicit and careful analysis is required of those that are specified as cost bearers and whose cost is specified, invisibly, as external cost.

It may be useful to differentiate between short-term and long-term perspectives. The probability of identifying scenario-dependent differences clearly increases as the time frame of the analysis expands. Furthermore, in the long term, complex estimates, the choice of system boundaries, and discounting rates usually have a dramatic influence on the results.

## 8.2 TEMPORAL ASPECTS AND DISCOUNTING OF LCC RESULTS

Conventional LCC applies market-based prices and discount rates (typically 4% to 15%) leading to a relevant time horizon of several decades at the most. Societal LCC uses alternative costs and social time preferences for discounting (a maximum of 0.1% and, in some cases, 1 to 2 orders of magnitude lower). This implies a time horizon of several thousand years. Strictly speaking, discounting is inconsistent with environmental LCC. Despite this, it is expected that most practitioners will apply it. One author (Hunkeler) believes that environmental LCC will use a higher discount rate in the short term with a Brundtland-like rate (Brundtland Commission 1987) following the economic life cycle of the product.

## 8.3 LEARNINGS FROM APPLIED LCC CARRIED OUT TO DATE

The applied studies in this book have been chosen because they all demonstrate a certain amount of vigor in carrying out the life cycle costing, as well as the complementary environmental analysis and/or sensitivity study. The cases demonstrate the avoidance of several pitfalls that risk the invalidation of systemic studies. Specifically, the 7 real and 1 hypothetical case studies presented in Chapter 7 are highlighted to illustrate the following points:

- Environmental LCC requires the use of the equivalent functional unit and system boundaries in both the economic and environmental assessments (see Section 7.2 for the case study on wastewater treatment).
- The portfolio representation — where LCC is plotted against a given, usually dominant, impact — should be carried out using absolute measures (e.g., monetary units versus global warming potential, as expressed in kg $CO_2$ equivalents).
- Portfolio representations are normalized, per se, per functional unit. Double normalization (e.g., per unit of GDP) can change even the ranking of alternatives. Therefore, any normalization other than by the functional unit bears the risk that the study would not hold up to peer review requirements.
- The majority of environmental LCC is based on the LCA approach and on standard LCA data packages and software.
- The most effective driver for LCC to be implemented within a firm is its top management. While LCM as a policy, or target, is widely advocated, the correlation between the successful realization of LCC thinking and the ability of the CEO to instill its importance is significant.
- In societal LCC, subsidies and, ultimately, tradable permits need to be monetized.
- In societal LCC taxes can be, in selected cases, larger than the monetized societal cost. Some would argue that a completely fair tax system would avoid the need for additional assessments on a given product category.
- In conventional LCC, which still dominates the literature, examples are seen where maintenance can dominate the LCC.

- The life cycle cost, as a ratio to the product selling price, can represent a small premium (e.g., Section 7.1, which contains the case study on extra-virgin olive oil) or a very large effect (e.g., Section 7.3 on illumination, and Section 7.4 on a train carriage), depending on the durability and effect of the use phase. The factors studies varied from 2 to 1000 in the cases presented in Chapter 7.
- In all types of LCC, the product perspective through the life cycle must be considered. At the very least, a gate-to-grave costing is required (in LCC, the assumption on upstream costs is that the market system adequately represents these).

## 8.4 STATE OF THE ART AND RULES OF THUMB IN CARRYING OUT LIFE CYCLE COSTING

A survey was carried out to identify use patterns in LCC. The 33 respondents from various geographical regions, industrial branches, and investigation units demonstrated, in particular for environmental and conventional LCC, deficiencies and strengths in the goal and scope definitions. Other questions that were unresolved relate to the tendency for those carrying out environmental LCC to rely, in the large majority, on cost estimation via prices, thereby neglecting parametric cost estimation techniques and other more advanced techniques, which are widely used for conventional LCC. Furthermore, while validation was often employed in conventional LCC, it seems less commonplace for environmental LCC. The following summary box provides some rules of thumb that *we* feel are guidelines for carrying out LCC.

**Rules of Thumb for Life Cycle Costing**

- State clearly all assumptions.
- Use a life cycle perspective (cradle-to-grave).
- Ensure that the functional unit describes functionality.
- Use real monetary flows, borne by some or more actors in the life cycle (for environmental and conventional LCC).
- Be transparent in terms of which costs are internalized (for societal LCC).
- Define thresholds appropriately.
- Discount the results (recommended for conventional and societal LCC).
- Validate the studies.
- Carry out a sensitivity analysis.
- Use relative life cycle portfolio presentations of LCC versus the main environmental impact.
- Benchmark against existing cases for similar products.
- Do not double normalize.
- Be concise in your discussion.
- Critical review is required for public assertions.

# 9 Outlook

## *Role of Environmental Life Cycle Costing in Sustainability Assessment*

*Walter Klöpffer*

## 9.1 SUSTAINABILITY

The term "sustainability" was introduced into the political, as well as public, discussion by the World Commission on Environmental and Development in the well-cited report *Our Common Future* (Brundtland Commission 1987). This document underlined the responsibility humankind has toward the future generations with an elegant definition that has had far-reaching acceptance from governments, NGOs, as well as private organizations:

> Sustainable development is development that meets the needs of present without compromising the ability of future generations to meet their own needs. (24 p)

Although this laudable claim was not easy to operationalize, it has been very successful in environmental politics as well as in mobilization. Indeed, the United Nations declared sustainability as the guiding principle for the 21st century at the World Conference in Rio de Janeiro and promoted a concrete action plan, Agenda 21 (United Nations Environment Programme [UNEP] 1992). The confirmation of this concept, in Johannesburg in 2002, introduced the life cycle aspect. Furthermore, the joint UNEP–SETAC Life Cycle Initiative was started just prior to the Johannesburg forum (Töpfer 2002). This initiative aims at a global promotion and use of life cycle thinking, life cycle assessment (LCA), and life cycle management (LCM).

Acting sustainability will require its quantification, the identification of appropriate and valid indicators, as well as associated thresholds. How this is achieved will be the topic of debate, though there is widespread belief that sustainability will involve an economic axis that will require life cycle costing. The standard model, which is well accepted by industry and often referred to as the triple bottom line, is a 3-pillar interpretation of sustainability. It states, essentially, that environmental, economic, and social aspects have to be tuned and checked against each another. One of the 1st uses of 3 dimensions in a life cycle method for products has been called "Produktlinienanalyse" (Öko-Institut 1987). This "product line analysis" was a proto-LCA (i.e., a life cycle assessment before the name, harmonization, and finally standardization of the different methods began approximately in 1990;

Klöpffer 2006). Product line analysis included an impact assessment, as the reader would refer to it today, with 3 dimensions — ecology, economy, and society. This clearly shows that the 3-pillar interpretation of sustainability is neither new nor an invention by industry. It is, therefore, rather straightforward to propose the following scheme in Equation (1) for sustainability assessment (*SustAss*) of products:

$$\text{SustAss} = \text{LCA} + \text{ELCC} + \text{SLCA} \tag{1}$$

LCA is the environmental life cycle assessment (SETAC 1993; ISO 14040/44 2006). ELCC stands for environmental life cycle costing (see Chapter 3), while SLCA stands for societal life cycle assessment. There are some prerequisites that have to be fulfilled in using Equation (1), the most important of which is that the system boundaries of the 3 assessments are consistent. This includes, of course, that in all 3 pillars of sustainability assessment the physical, as opposed to the marketing, life cycle is used for the life cycle inventory (LCI; ISO 14040/44 2006). The 3 pillars may even be able to utilize the same inventory data, using the language of ISO 14040, with the caveat being that SLCA may require the introduction of geographically specific data (Hunkeler 2006).

The underlying driver as to why sustainability assessment methods (ELCC and SLCA) have to be life cycle based is rather easily explained (Klöpffer 2003):

> Only in this way, trade-offs can be recognized and avoided. Life cycle thinking is the prerequisite of any sound sustainability assessment. It does not make any sense at all to improve (environmentally, economically, socially) one part of the system in one country, in one step of the life cycle or in one environmental compartment, if this "improvement" has negative consequences for other parts of the system which may outweigh the advantages achieved. Furthermore, the problems shall not be shifted into the future.

The last point, avoiding the shifting of problems into the future, is of paramount importance due to the request for intergenerational justice (Brundtland Commission 1987). Life cycle thinking alone is not enough, however, since in order to estimate the magnitude of the trade-offs, which are nearly always present, the instruments required have to be as quantitative as possible. Since we are living in a global economy, the system boundaries used in the methods must also be global. Within this context, the UNEP–SETAC life cycle initiative deserves attention and support.

## 9.2 STATUS OF DEVELOPMENT

### 9.2.1 LIFE CYCLE ASSESSMENT (LCA)

LCA is the only internationally standardized environmental assessment method (ISO 14040 series). The historical development of LCA, beginning with the proto-LCAs of the 1970s and 1980s, such as Hunt and Franklin (1974) or Ökoinstitut (1987), has been summarized (Klöpffer 2006). The international standards have recently been slightly revised with the modified standards ISO 14040/44 2006 in October 2006 (Finkbeiner et al. 2006) superseding the older series ISO 14040 (1997), 14041 (1998), 14042 (2000a), T9043 (2000b). On the other hand, it is well known that

LCA is an active research field, so further methodological developments are to be expected. A recent textbook on LCA outlines the development as well as the method and the most important applications (Baumann and Tillman 2004). The Dutch LCA guidelines can be considered a comprehensive recent monograph, based on the ISO series of LCA standards (Guinée et al. 2002). A similar endeavor from the Danish point of view dates back to the late 1990s (Wenzel et al. 1997; Hauschild and Wenzel 1998). The basic principles of LCA, which together distinguish this method from other environmental assessment methods, are as follows:

- The analysis is conducted "from cradle to grave."
- All mass and energy flows, resource and land use, as well as potential impacts connected with these "interventions" are set in relation to a functional unit as a quantitative measure of the benefit of the system(s).
- The method is comparative; LCAs are restricted to improvements of only 1 system, even if the future state is compared to the present one.

The advantage, at least theoretically, of the completeness is partly offset by the uncertainty regarding where and when exactly some of the processes or emissions occur, which ecosystems or how many humans may be harmed, and whether or not thresholds of effects are really surpassed due to the emissions or other effects that can be attributed to the systems studied. Furthermore, the magnitude of the reference flows, which quantify the functional unit, is usually fixed arbitrarily. As a consequence, the absolute amount of the "interventions" (i.e., emissions and use of resources) has no meaning, and concentrations of emitted substances cannot be calculated (hence, no risk assessment is possible). The additional use of other absolute, noncomparative methods (e.g., risk assessment, material, and substance flow analysis) is, therefore, recommended for the sake of decision making. It is difficult, however, to integrate such additional methods directly into LCA studies. This may be seen as a disadvantage, though it is outweighed by the advantages of standardization of LCA, such as a clear structure and measures against misuse, for example in marketing. Sonnemann et al. (2003) describe, for industrial processes, the integration of LCA and environmental risk assessment.

The ISO structure of LCA goes back to a very similar scheme proposed by SETAC (1993) and now consists of the following four components (ISO 14040/44 2006):

- Goal and scope definition
- Inventory analysis
- Impact assessment
- Interpretation

If comparative assertions (e.g., system A is better than or equal to system B in regard to its environmental aspects) are part of an LCA and are intended to be made available to the public, a critical review is mandatory according to the panel method (i.e., by at least 3 reviewers; ISO 14040/44 2006). This and many other "obstacles" were built into the ISO series of LCA standards in order to prevent their misuse, especially by false public claims. As a consequence of these preventive measures, a full LCA to be used publicly has become a somewhat lengthy procedure. Of course, the learning process is more rewarding in a long and carefully conducted LCA study compared to a "quick and dirty" one.

It has been widely observed that the design phase permits, quite often, only cursory assessments, and, therefore, simplified methods were proposed and also compared with each other (Hunt et al. 1998). In design for environment, a compromise has to be found between a reasonably comprehensive coverage of the life cycle, the impact categories, and the time needed for data collection and modeling. The actual calculation process is rapid due to the elaborate LCA software that is now available. It is also true that additional and higher quality data have become available recently (Frischknecht et al. 2005). It should also be noted that the standards are much more flexible and less demanding if the results are used internally. In this case, the critical review is optional and can be performed by a single internal or external expert instead of a panel according to ISO 14040/44 (2006), and weighing between results of different impact categories is allowed.

### 9.2.2 Life Cycle Costing (LCC)

LCC is older than LCA, though it is not yet standardized. It also has a large potential for extending the scope of LCA in the direction of sustainability assessment (Hunkeler and Rebitzer 2001; Norris 2001; Rebitzer 2002; Klöpffer 2003; Klöpffer and Renner 2008). This environmental LCC (see Chapter 3) is based on the physical life cycle used in LCA and avoids the monetization of externalities that are not to be internalized in the decision-relevant future, since this would mean a double counting: environmental impacts are quantified in the life cycle impact assessment (LCIA) component of LCA in physical units (ISO 14040/44 2006).

It should be noted that LCC includes the use and end-of-life phases (i.e., from cradle to grave, as in LCA), so that the result cannot be approximated by the price of a product (so-called cradle to factory gate or cradle to point of sale). Furthermore, LCC is an assessment method, not an economic cost-accounting method. Potential links of sustainability assessment and, in particular, LCC with environmental management accounting have been discussed by Klöpffer and Renner (2007).

### 9.2.3 Societal Life Cycle Assessment (SLCA)

SLCA is generally considered to be still in its infancy, although the idea is not new (Öko-Institut 1987; O'Brian et al. 1996). Quite to the contrary, a significant increase in the number of papers published can recently be observed:

- Dreyer et al. (2006) aim at assessing the responsibility of the companies involved, although the products are the point of reference. This necessarily gives more weight to the foreground activities and to the people involved in them.
- Labuschagne and Brent (2006) strive for completeness of the social indicators to be used in a social impact assessment.
- Weidema (2006) includes elements of cost–benefit analysis (CBA) and proposes quality-adjusted life years (QALY) as a main measure of human

health and well-being (a common endpoint for toxic and social health impacts). The author holds the view that social impacts should be treated within LCA as a special section of impact assessment, that is, a common inventory (LCI) would be required.

- Norris (2006) considers social and socioeconomic impacts leading to poor health; life cycle attribute assessment as a web-based instrument should complement classical LCA methods.
- Hunkeler (2006) deals with the connection of societal indicators with the functional unit. This is a daunting problem, given the mostly qualitative nature of societal indicators and the need for quantification in comparative assessments. It seems to now be near a solution, taking the working hours spent per functional unit as the link; furthermore, regional income per hour and the number of working hours needed to satisfy important social needs (e.g., education and health care) are used to quantify the different social development statuses of the regions. The higher regional resolution needed for the establishment of societal life cycle inventories (SLCIs) will be a challenge for the LCA community, but, on the other hand, there are researchers claiming for a much better regional resolution in LCA or LCIA, too (Potting and Hauschild 2006).

## 9.3 DISCUSSION

There are at least 2 options to include the social aspects into a life cycle–based sustainability assessment. The 1st option corresponds to Equation (1) and is based on 3 separate life cycle assessments with consistent system boundaries and the same functional unit (Klöpffer 2003). A formal weighting between the 3 pillars, although possible, would, more appropriately, be avoided. The main advantage of this approach is its transparency. The attribution of advantages and disadvantages in comparative assessments is clear in this variant; there is no compensation between the pillars. As a consequence, a favorable (economic) ELCC result for a given product cannot outweigh less favorable or even poor results in (environmental) LCA and SLCA. Such an overweighting of the economic part, as is daily practice in business today, would perpetuate the (largely unsustainable) status quo of economy.

SustAss = LCA (new, expanded relative to ISO 14040/44 [2006], and including elements of ELCC and SLCA as additional impacts in LCIA). (2)

The 2nd option (equation (2)) would imply that 1 LCI would be followed up by 3 impact assessments covering the potential environmental, economic, and social impacts per functional unit of the product system studied. The advantage of this option would be that the same LCI could be used for all 3 impact assessments, solving the system boundary problem. Such a solution seems to be preferred by Weidema (2006). Disregarding for the moment the danger of mixing up the 3 dimensions, there remains the question of whether or not option 2 is compatible with ISO.

The revised framework standard ISO 14040/44 (2006) says, "LCA addresses the environmental aspects and potential impacts" throughout a product's life cycle from raw material acquisition to production, use, end-of-life treatment, recycling, and final disposal (i.e., from cradle to grave). It also noted, "LCA typically does not address the economic or social aspects of a product, but the life cycle approach and methodologies described in this International Standard may be applied to these other aspects."

These statements clearly favor option 1, and future separate standardizations of ELCC and SLCA would be a logical consequence. On the other hand, ISO 14040/44 (2006) could be revised again in the future and possibly accommodate economic and societal impact assessments. Since this revision will certainly not begin in the immediate future, the coming years should be used for discussing the best way to formalize sustainability assessment. Additional experience with the new social indicators will be required, in particular establishing the means to unambiguously link them to the functional unit of a product system. The selection and quantification of the most appropriate metrics per functional unit will be the main scientific problem regardless of whether option 1 or 2 will be followed. As in LCIA, it will not be possible to properly quantify all desirable impacts.

# Appendices

## APPENDIX TO CASE STUDY BOXES: WASHING MACHINES AND PRIVATE LAUNDRY IN EUROPE, NORTH AMERICA, AND ASIA AND JAPAN

Case study boxes on a washing machine are presented throughout this book to illustrate conventional, environmental, and societal LCC. Table 7.8 provides an overview of all washing machine case study boxes.

Nevertheless, the North American reader may be surprised by some of the washing machine technologies and consumer habits presented, as almost all of the case study boxes are based on European cultural practices of laundering, influencing the material and production technology chosen and the use phase modeled. Beyond, an input–output analysis of Japanese washing machines was carried out for Case Study Box 8, in comparison to the LCC for European washing machines.

The authors consider it essential to summarize in the following the usual differences between European, Japanese, and North American washing machines, which the reader should keep in mind when studying the common example used throughout the book or when transferring the example to North American or other conditions. Cultural practices for laundering, and therefore the hardware, vary in different regions, and thus studies across regions cannot be directly compared without recognizing this fact.

There are two basic washing machine designs: vertical axis machines and horizontal axis drum devices. In North America and Asia the vertical axis type dominates, whereas this kind of machine has been entirely replaced in Europe by the horizontal axis device. In North America, vertical axis machines with a central "agitator" are used; in Asia, impeller machines with a ribbed disk mounted at the bottom of the tub are used. Vertical axis machines are loaded from the top, and horizontal axis machines mostly from the front (there are also so-called top loaders with a horizontal axis; however, these only have a quite small market share). Recently, however (i.e., since the 1990s), horizontal axis machines were also introduced in non-European markets, mainly due to their better water and energy efficiency (Smulders 2002).

In contrast to horizontal axis machines, vertical axis machines usually do not have internal heating, though they are connected to both a hot and a cold water tap. Another difference is the weight: horizontal axis machines need a critical weight made from concrete or steel for stability reasons. Therefore, European washing machines are much heavier than American or Asian ones (70 to 100 kg compared to approximately 30 kg). The main advantage of the vertical axis machines is that they wash much faster than horizontal axis machines. The average duration of a European washing cycle is 90 minutes, whereas in Japan it takes 60 minutes and in North America only 35 minutes. However, vertical axis machines have a higher

**TABLE A.1**
**Specifications of washing machine types used in Europe, North America, and Asia and Japan**

| | Europe | North America | Asia/Japan | Reference |
|---|---|---|---|---|
| Machine type | Horizontal axis | Vertical axis, agitator type | Vertical axis, impeller type | Smulders (2002) |
| Capacity | 5 to 7 kg | 5 kg | 3 to 8 kg | Rüdenauer et al. (2004) and Smulders (2002) |
| Internal heating circuit | Yes | No | No | Smulders (2002) |
| Weight | 70 to 100 kg | 30 kg (own estimation) | 30 kg (Japan) | Rüdenauer et al. (2004) and Matsuno et al. (1996) |
| Cycle length | 90 minutes | 35 minutes | 60 minutes (Japan) | Metzger-Groom (2003) |
| Water consumption (without rinsing cycles) | 4 l/kg laundry | 25 l/kg laundry | 10 to 15 l/kg laundry | Smulders (2002) |

water demand compared to horizontal axis machines: North American agitator-type vertical axis machines need, for example, some 25 liters of water (without rinsing) per kg of laundry; Asian machines need 15 to 20 liters; and European horizontal-type machines need only some 4 liters (depending on the wash program; see Table A.1).

Next to these machine specifications, washing habits also differ in these regions (all data from Metzger-Groom 2003): the average wash temperatures in Europe are much higher than those in North America and Japan (42°C in Europe versus 29°C and 23°C respectively). Additionally, European households use more detergent than North Americans and Japanese (120 grams per wash load compared to 60 and 30 grams respectively). These 2 factors result in a much better cleaning performance in Europe: the cleaning performance in North America is less than 80%, and the Japanese cleaning performance only about 65%, of the European level.

The higher temperatures and larger amounts of detergents, however, are somewhat compensated for by less washes per week: in general, European households wash 5 times per week, North American households 7 times per week, and Japanese households even 10 times per week. Further differences concern, for example, pre-treating of the garment or the usage of bleaching agents.

## APPENDIX TO CHAPTER 4: SOCIAL IMPACTS

Table A.2 summarizes the various social impacts, noting also the potential relevance for life cycle costing.

**TABLE A.2**
**Social impacts and their relevance for LCC**

| Social impact | Relevance for LCC (example) | Comments |
|---|---|---|
| | **Health and social well-being** | |
| Death (of self, in family, or in the community) | Products with a direct fatal impact (weapons), accidents due to products, or the like | Could be related to statistical number of fatalities |
| Reduced number of fatalities in society | Safety product features (e.g., airbags and pedestrian protection) | Could be related to statistical number of reduced fatalities |
| Nutrition | Products improving nutrition (e.g., fertilizer, food packaging, and refrigerants) or poisoning impacts during the life cycle | Could be related to statistical numbers of changed yield per acre |
| Actual physical or mental health and fertility (reduced or improved by product impact) | Pharmaceutical products or negative impacts during the life cycle | Could be related to statistical numbers of illness impacts |
| Perceived health | Placebos (e.g., from electromagnetic pollution) | Percentage of population suffering from diffuse health impacts |
| Aspirations and image | Luxury products | Market analysis |
| Autonomy | Products enabling individual mobility, communication, and so on | |
| Stigmatization or deviance labeling | Energy-efficient appliances | |
| Feelings in relation to the project | Big infrastructural projects | Survey |
| | **Quality of the living environment (livability)** | |
| Quality of the living environment (actual and perceived) | Similar issues that are treated in environmental impact assessments | Avoid double counting with LCA |
| Leisure and recreational opportunities and facilities | Landscape-changing and land-consuming products | Avoid double counting with LCA |
| Environmental amenity value and/or aesthetic quality | Landscape-changing and land-consuming products | Avoid double counting with LCA |
| Availability of housing facilities, physical quality of housing (actual and perceived), and social quality of housing (homeliness) | Housing products | Affordability and quality aspects |
| Adequacy of physical infrastructure | Communication and mobility products and services | Distance to target or average relation between population and infrastructure |

*(continued)*

**TABLE A.2**
**Social impacts and their relevance for LCC (continued)**

| Social impact | Relevance for LCC (example) | Comments |
|---|---|---|
| Adequacy of and access to social infrastructure | Health care products | Health costs and the like |
| Personal safety and hazard exposure (actual and perceived) | Hazardous chemicals or waste in the life cycle | Could be related to statistical number of accidents |
| Crime and violence (actual and perceived) | Security products and indirect impacts along the life cycle | Could be related to statistical numbers of crime and violence |
| | **Cultural impacts** | |
| Change in cultural values (moral rules, beliefs, etc.), or cultural affront | Products in conflict with cultural values in different regions | — |
| Cultural integrity | Media products | — |
| Experience of being culturally marginalized | Roads in areas with indigenous populations | — |
| Profanation of culture | Media products | — |
| Loss of language or dialect | Products standardizing a certain language (software) | Qualitatively |
| Natural and cultural heritage (violation, damage, or destruction) | Infrastructural projects | Avoid double counting with LCA |
| | **Family and community impacts** | |
| Alteration of family structure | Linked to life cycle impacts of projects or products (e.g., by job losses) | — |
| Obligations to living family members and ancestors<br>Family violence<br>Social networks<br>Community identification and connection | Unlikely to be monetized and more reasonably expressed as a separate set of midpoint indicators | To be included in a complementary societal assessment |
| Community cohesion (actual and perceived)<br>Social differentiation and inequity<br>Social tension and violence | Unlikely to be monetized and more reasonably expressed as a separate set of midpoint indicators | To be included in a complementary societal assessment |
| | **Institutional, legal, political, and equity impacts** | |
| Functioning of government agencies | Government projects | Could be related to changes in time needed for bureaucratic activities |
| Access to legal procedures and legal advice | Unlikely to be monetized and more reasonably expressed as a separate set of midpoint indicators | — |
| Integrity of government and government agencies | | — |

**TABLE A.2**
**Social impacts and their relevance for LCC (continued)**

| Social impact | Relevance for LCC (example) | Comments |
|---|---|---|
| Participation in decision making | Government projects | Could be related to % of participation |
| Tenure or legal rights | Products and projects related to data safety | To be captured qualitatively |
| Subsidiary (the principle that decisions should be made as close to the people as possible) | N/A, or see "Participation in decision making," above | — |
| Human rights | Often captured by other social impacts | — |
| Impact equity | See below | — |
| **Relations between people with different genders, ethnicities, races, ages, sexual orientations, religions, opinions, education levels, income levels, presence of disabilities, and so on** | | |
| Physical integrity | Products with encouraging or discouraging features or information | Specific ways for measurement (e.g., psychological analysis) |
| Personal autonomy | Unlikely to be monetized and more reasonably expressed as a separate set of midpoint indicators | — |
| Fair division of production-oriented labor | Products or projects enabling work for different groups (part-time, or kindergarten) or impacts along the product life cycle | Could be related to changes in % of labor |
| Fair division of household labor | Unlikely to be monetized and more reasonably expressed as a separate set of midpoint indicators | — |
| Fair division of reproductive labor | Impacts along the product life cycle | Percentage of participation for each group |
| Fair control over and access to resources | Fair trade products | — |
| Equal political emancipation | N/A | — |
| Equal access to services (mobility, communication, health care, etc.) | Product features enabling use of, for example, mobility carriers by disabled people | Specific measures (e.g., wheelchair versus vehicle dimension) |

*Source:* Based on Schooten et al. (2003).

## APPENDIX TO CHAPTER 6: SURVEY FORM: FOR INVESTIGATION OF LCC PARAMETERS

Andreas Ciroth and Christian Trescher

With contributions from Wulf-Peter Schmidt, Andrea Heilmann, Gerald Rebitzer, David Hunkeler, and others, for use within the SETAC-Europe working group on Life Cycle Costing

### MOTIVATION

For performing LCC studies, numerous goal and scope settings are possible that shall, ideally, be reflected in the approach and methods used in the studies.

For further analyzing this brief idea, the following text investigates

- different goal and scope settings and
- different methods and methodological choices used in LCC studies.

First, the goal is to empirically investigate different method–goal combinations (i.e., which combinations take place in existing case studies of the present and past?). This step could be labeled a descriptive step. Second, the goal is to come to recommendations for performing LCC studies (i.e., to derive implications of different goal and scope settings on the selected methods).

> For a start, the authors have the aim to fill the following form with examples from case studies.

### A. Goal, Scope, and Background

1) Reason for performing the study (decision to be supported; who, or which event, gave the reason; are there different parties to be distinguished; is there a general regulation that promotes it; asf.)

2) Source, reference for the study

3) Study performed by
   a. ☐ External contractor (consultant)
   b. ☐ Internal sources
      If a or b: share between both (external 0% to 100%?)

   c. Date and country of study

4) Intended use
   a. Type of use: individual case, update, controlling, performance evaluation, tender, and/or other (description)
   b. Is the study done after, parallel to, or independent from other LCM* assessments (e.g., environment or social)? If so, which?

5) Which types of branch was or were touched?

6) What was the object of study?
   a. Description
   b. Functional unit**

7) Costs of the considered object
   a. Overall LC costs as given in study

   b. Relation of investment costs or purchase costs to the overall LC costs (purchase costs for the virgin product, and first sale)

8) Does the life cycle considered span different countries, does it integrate costs from different sources? If so, which (sources may be listed per type: businesses, statistics, market information, others)?

9) Time frame
   a. Duration of study (may be differentiated between initial motivation for performing the study, kickoff, finish)
   b. Time span covered by life cycle

10) Addressees of study
   a. ☐ Internal (management or other)
   b. ☐ External (client, supplier, bank, other involved in companies' business)
   c. ☐ External (public, other specific audience not involved in companies' business)
   d. Specific definition of the decision maker (who makes the decision)

   e. Sphere of influence of the decision maker (i.e., what parts of the LC can be influenced by the decision?)

   f. List of stakeholders involved and their roles

---

* LCM: Life Cycle Management

** A "functional unit" is the unit of the object of study, for which the study is performed and the LC costs are provided, described as precisely as possible (e.g., 100 light bulbs, 60W, clear glass, standard, non–energy-saving type).

11) Relevance
   a. ☐ Study (method development, primarily case study for applying newly developed methods)
   b. ☐ Practical decision support (short-term consequences only)
   c. ☐ Practical decision support (decision with long-term, contractual consequences)

## B. Result

12) Type of costs considered (investment, maintenance, etc.)

13) Type of costs not considered (investment, maintenance, etc.)
   a. Remarks: Was the consideration of cost types steered by intention or by other reasons?

14) Parts of the LC excluded (single LC stages like production, use, maintenance, repair in use stage, recycling, final disposal)

15) ☐ Uncertainty consideration in result?
   a. If yes: relative amount of uncertainty in result as given?*

16) Other aspects of object considered and investigated (reliability, energy consumption, etc.)

17) Internal costs alone or also external costs** considered? Which type of external costs, if applicable?

## C. Approach

18) Source of approach
   a. ☐ Consultant
   b. ☐ Consultant and client
   c. ☐ Generic

19) Description of approach and main assumptions

20) Discounting rate as applied (0, if no discounting is applied in study)

21) Description of different scenarios investigated, if applicable

---

* For a question on the approach for uncertainty estimation, see item 24c, below (regarding approach).

** External costs of a product represent the monetized effects of environmental and social impacts related to the product. External costs are, in contrast to internal costs, not directly borne by the firm, consumer, government, or the like that is producing, using, or handling the product (modified from Rebitzer and Hunkeler 2003).

22) Special approaches
   a. ☐ Simulation performed
   b. ☐ Prognosis performed
   c. ☐ Uncertainty in cost data considered? If so, how?

   d. ☐ Long-term data collection performed or to be performed?

23) Data sources
   a. ☐ Internal (company, nonpublic)
   b. ☐ External (e.g., market information, public statistics, literature)
   c. ☐ Expert judgment

24) Approach of cost estimation used
   a. ☐ Price
   b. ☐ Parametric*
   c. ☐ Via functional relations (other than parametric)
   d. ☐ Others:

25) Software used (and for which purpose)
   a. ☐ HPP (hand, pencil, and paper), for purpose (data collection, analysis, simulation, and prognosis):
   b. ☐ Spreadsheet, for purpose:
   c. ☐ Database, for purpose:
   d. ☐ LCC or TCO tool, for purpose:
   e. ☐ Other, for purpose:

## D. Additional Remarks

---

* "Parametric Cost Estimating" — a cost estimating methodology using statistical relationships between historical costs and other program variables such as system physical or performance characteristics, contractor output measures, and manpower loading. An estimating technique that employs one or more cost estimating relationships (CERs) for the measurement of costs associated with the development, manufacture, and/or modification of a specified end item based on its technical, physical, or other characteristics" (US Department of Defense 1999).

# Glossary

**conventional LCC:** An assessment of all costs associated with the life cycle of a product that are directly covered by any 1 or more of the actors in the life cycle.

**cost:** The cash or cash equivalent value sacrificed for goods and services that are expected to bring a current or future benefit to the organization (Hansen and Mowen 1997).

**cost management:** Encompasses all (control) measures that aim to influence cost structures and cost behavior precociously. Among these tasks, the costs within the value chain have to be assessed, planned, controlled, and evaluated (Dellmann and Franz 1994). A cost management system is a set of cost management techniques that function together to support the organization's goals and activities (Hilton et al. 2000).

**discounted cash flow:** By discounting the future cash flow (i.e., using an interest rate that reflects the fact that money in the future is worth less than money now), one can calculate, for example, net present and net future values. The interest rate is a means of reflecting the opportunity costs of tying up money in the investment project (from Economist.com 2007).

**discounting:** Converts costs (and revenues or value) occurring at different times to equivalent (net) costs at a common point in time.

**environmental cost:** This has 2 basic definitions:

1) Environmental damage expressed in monetary terms = cost of externalities/ external effects.
2) The market-based cost of measures to prevent environmental damage, including EoL processes. Market-based costs are part of life cycle costing.

**environmental LCC:** An assessment of all costs associated with the life cycle of a product that are directly covered by any one or more of the actors in the product life cycle (e.g., supplier, manufacturer, user or consumer, or EoL actor) with complementary inclusion of externalities that are anticipated to be internalized in the decision-relevant future. (Definition as suggested by Rebitzer and Hunkeler 2003.) Environmental LCC has to be accompanied by a life cycle assessment and is a consistent pillar of sustainability.

**EoL processes:** End-of-life processes comprise all processes after the use phase in the life cycle of a product; hence collection, disassembly, re-use, recycling, composting, landfill; and/or incineration.

**external cost:** This has 2 different meanings:

1) Cost of externalities, as welfare effects. Being nonmarket effects, they are measured by other means, as through surveys on willingness to pay.
2) Cost, as market cost, not directly borne by an organization in terms of costs of labor, capital, and taxes, but as costs for purchases from other firms in the system, covering the internal costs of these other firms.

**external effect (or the externality effect):** The effect of an economic activity on the welfare of individuals that is not reflected in the prices in the markets related to this activity. Most economists focus nowadays on externalities on environmental externalities, but other externalities may be distinguished, like the effects of knowledge created by schooling and research on the welfare of others.

**externalities:** Value changes caused by a business transaction but not included in its price.

**financial accounting information system:** An accounting information subsystem that is primarily concerned with producing outputs for external users and uses well-specified economic events as inputs and processes that meet certain rules and conventions (Hansen and Mowen 1997).

**internal cost:** Cost directly borne by an individual or organization in supplying or consuming a product, as value added by the firm (capital and labor costs). Complement of external cost (definition 2).

**life cycle (LC):** All processes or activities involved in having a unit of function of a product, including all life cycle stages, from primary materials production and manufacturing through use to final disposal activities (physical life cycle concept).

**life cycle management (LCM):** An integrated framework of concepts and techniques to address environmental, economic, technological, and social aspects of products, services, and organizations. LCM, like any other management pattern, is applied on a voluntary basis and can be adapted to the specific needs and characteristics of individual organizations (Hunkeler et al. 2004).

**monetary externalities:** Externalities expressed in monetary terms.

**product:** Material good or service; also, a commodity. The terms "product and commodity" sometimes refer to material goods only.

**revenues:** Inflows or other enhancements of assets of an entity, settlements of an entity's liabilities, or a combination of both from delivering or producing goods, rendering services, or engaging in other activities that constitute the entity's ongoing major or central operations (Financial Accounting Standards Board n.d.).

**societal LCC:** An assessment of all costs, including costs of externalities, associated with the life cycle of a product that are covered by any actor in society. Transfer payments are not considered in societal LCC.

**transfer payments:** Payments between governments and private persons or organizations, involving taxes and subsidies. Payments for public services, like for waste management, may fall under this heading if paid (for example) by a local municipality from taxes or levies.

**value added:** The difference between the cost of products purchased and the proceeds of products sold, as gross value added, being the costs of labor and capital, including profits. Net value added is obtained by subtracting depreciation from gross value added.

# References


Ackerman, F. 2004. Priceless: human health, the environment and limits of the market. New York (NY): New Press. 288 p.

Allenby, B, Yasui, I, Lehni, M, Zust, R, Hunkeler, D. 1998. Ecometrics' stakeholder subjectivity. Env. Quality Mgt. (Autumn):1.

Allgemeiner Deutscher Automobil (ADAC), ed. 2005. ADAC Spezial Autokosten (CD-ROM). 2nd ed. Munich (Germany): ADAC.

American Society for Testing and Materials (ASTM). 1994. Standard practice for environmental site assessments: E 1527 - 94. West Conshohocken (PA): American Society for Testing and Materials.

AQUA+TECH. n.d. AQUA+TECH. http://www.aquaplustech.com.

Associazione Italiana per l'Agricoltura Biologica. 2007. AIAB guidelines. http://www.aiab.it.

Australian Department of Defence. 1998. Life-cycle costing in the Department of Defence. Canberra (Australia): Department of Defence, Australian National Audit Office.

Badgett, L, Hawke, B, Humphrey, K. 1995. Analysis of pollution prevention investments using total cost assessment: a case study in the electronics industry. Seattle (WA): Pacific Northwest Pollution Prevention Research Center.

Bage, G, Samson, R. 2003. The econo-environmental return (EER): a link between environmental impacts and economic aspects in a life cycle thinking perspective. International Journal of Life Cycle Assessment 8(4):246–251.

Bahnkreis. 2000. Verbundprojekt, Förderkennzeichen des BMBF: 02PV21319, Band 3 Baugruppenmodell — Ein Instrument zur ökologischen und ökonomischen Beurteilung von Schienenfahrzeugbaugruppen, Fleischer G, Ciroth A, Gerner K, Kunst H. Berlin (Germany): Bahnkreis. http://edok01.tib.uni-hannover.de/edoks/e01fb01/330464833.pdf.

Bartolomeo, M, Bennett, M, Bouma, JJ, Hetdkamp, P, James, P, Wolters, T. 2000. Environmental management accounting in Europe: current practice and future potential. European Accounting Review 9(1):31–52.

Bateman, IJ, Willis, K. 2002. Valuing environmental preferences: theory and practice of the contingent valuation method in the US, EU and developing countries. Oxford (UK): Oxford University Press.

Baumann, H, Tillman, AM. 2004. The hitch hiker's guide to LCA: an orientation in LCA methodology and application. Lund (Sweden): Studentlitteratur.

Becker, HA, Vanclay, F, editors. 2003. The international handbook of social impact assessment. Cheltenham (UK): Edward Elgar. 326 p.

Bickel, P, Friedrich, R. 2005. ExternE externalities of energy: methodology 2005 update, European Communities. http://www.externe.info.

Biswas, G, Clift, R, Ehrenfeld, J, Forster, R, Jolliet, O, Knoepfel, I, Luterbacher, U, Russell, D, Hunkeler, D. 1998. Ecometrics: identification, characterization and life cycle validation. International Journal of Life Cycle Assessment 3:184.

Blanchard, B. 1978. Design and manage to life cycle cost. Portland (ME): M/A Press.

Blanchard, B, Fabrycky, WJ. 1998. Systems engineering and analysis. Upper Saddle River (NJ): Prentice Hall. 506 p.

Boehm, B. 1981. Software engineering economics. Englewood Cliffs (NJ): Prentice Hall.

Bonz, M. 1997. Lebens-Zykluskosten als Entscheidungskriterium für die Beschaffung von Fahrzeugen und Anlagen, Life-cycle costs as criterion for the purchase of public transport vehicles. In: VDI Congress, editor. Systemoptimierung im spurgeführten Verkehr. Munich, September 25–26, proceedings. p 7–14 (in German).

Braune A. 2002. Life cycle Modellierung von Abwasserreinigungsoptionen. Swiss Federal Institute of Technology Lausanne. Diploma thesis, Berlin Technical University.

Brundtland Commission. 1987. Our common future. Oxford (UK): World Commission on Environment and Development.

Ciroth, A. 2002. The time dimension in an LCA for a train's component, "trainEE." Presentation LCA forum, Zurich.

Ciroth, A, Gerner, K, Ackermann, R, Fleischer, G. 2003. IT-Lösungen für den Bahnkreis — Datenbank- und Softwareentwicklung zur Darstellung der Umweltrelevanz von Schienenfahrzeugen, Handbuch Umweltwissenschaften. Lampertheim (Germany): Alpha. p 95–102.

Ciroth, A, Trescher, C. 2004. Survey form proposed for use in the SETAC LCC working group. October 2003, Version 2–Input Form, Berlin, Eltville (Germany).

Cooper, RG. 2001. Winning at new products: accelerating the process from idea to launch. Cambridge (MA): Perseus.

Culham, E. 2000. Economic evaluation of PVC waste management. Report for European Commission Environment Directorate. Abingdon (UK): AEA Technology.

Dasgupta, AK, Pearce, DW. 1972. Cost-benefit analysis: theory and practice. London (UK): Macmillan.

Dasgupta, P, Marglin, S, Sen, AK. 1972. Guidelines for project evaluation. Vienna (Austria): United Nations Industrial Development Organisation (UNIDO).

Dellmann, K, Franz, K-P. 1994. Von der Kostenrechnung zum Kostenmanagement. In: Dellmann, K, Franz, K-P, editors. Neuere Entwicklungen im Kostenmanagement. Stuttgart (Germany): Verlag Paul Haupt. p 15–30.

Dhillon, BS. 1989. Life cycle costing: techniques, models and applications. London (UK): Taylor & Francis.

Dorfman, R, Samuelson, P, Solow, R. 1958. Linear programming and economic analysis. New York (NY): McGraw-Hill.

Dreyer, LC, Hauschild, MZ, Schierbeck, J. 2006. A framework for social life cycle impact assessment. International Journal of Life Cycle Assessment 11(2):88–97.

Ecole Polytechnique Fédérale de Lausanne. 2007. News. http://www.epfl.ch.

Economist.com. 2007. Economics a to z. http://www.economist.com/research/Economics.

EcoTopTen. 2007. New ecology products. http://www.ecotopten.de.

Ekvall, T, Weidema, B. 2004. System boundaries and input data in consequential life cycle inventory analysis. Journal of Life Cycle Assessment 9(3):161–171.

Ellram, LM. 1993. A framework for total cost of ownership. International Journal of Logistics Management 4(2):49–60.

Ellram, LM. 1994. A taxonomy of total cost of ownership models. Journal of Business Logistics 15(1):171–192.

Ellram, LM. 1995. Activity-based costing and total cost of ownership: a critical linkage. Journal of Cost Management 9(4):22–30.

Endres, A. 1982. Ökonomische Grundprobleme der Messung sozialer Kosten. List Forum, 11(4):251–269.

EU DG RTD. 1995. Metholodogy. Vol. 2 of ExternE project. Brussels (Belgium): EU DG RTD.

EU DG RTD. 1999a. Assessment of global warming damages. Vol. 8 of ExternE project. Brussels (Belgium): EU DG RTD.

EU DG RTD. 1999b. ExternE: externalities of energy: Methodology 1998 update. Brussels (Belgium): EU DG RTD.

European Union. 2001. Green paper on integrated product policy. Brussels (Belgium): Commission of the European Communities.

European Union. 2003a. Directive 2002/96/EC of the European Parliament and of the Council on waste electrical and electronic equipment (WEEE). Brussels (Belgium): EU Commission and Parliament.

European Union. 2003b. Towards a thematic strategy on the sustainable use of natural resources. Brussels (Belgium): Commission of the European Communities.

European Union. 2005a. Directive 2005/32/EC of the European Parliament and of the Council on establishing a framework for the setting of ecodesign requirements for energy-using products. Brussels (Belgium): Commission of the European Communities.

European Union. 2005b. Impact assessment guidelines, SEC(2005) 791, June 2005 with March 2006 update. Brussels: Commission of the European Communities.

Fava, J, Consoli, F, Denison, R, Dickson, K, Mohin, T, Vigon, B, editors. 1993. A conceptual framework for life cycle impact assessment. Pensacola (FL): SETAC Press.

FEEM. 1997. ExternE national implementation Italy. Final report. Rome (Italy): FEEM.

Financial Accounting Standards Board. N.d. US accounting standards. Norwalk (CT): Financial Accounting Standards Board.

Finkbeiner, M, Inaba, A, Tan, RBH, Christiansen, K, Klüppel, H-J. 2006. The new international standards for life cycle assessment: ISO 14040 and ISO 14044. International Journal of Life Cycle Assessment 11(2):80–85.

Fleischer, G, Ciroth, A, Gerner, K, Kunst, H. 2000. Verbundprojekt Bahnkreis, Förderkennzeichen des BMBF: 02PV21319, Band 3 Baugruppenmodell — Ein Instrument zur ökologischen und ökonomischen Beurteilung von Schienenfahrzeugbaugruppen. Berlin (Germany): Verbundprojekt Bahnkreis.

Frischknecht, R, Jungbluth, N, Althaus, H-J, Doka, G, Dones, R, Heck, T, Hellweg, S, Hischier, R, Nemecek, T, Rebitzer, G, Spielmann, M. 2005. The ecoinvent database: overview and methodological framework. International Journal of Life Cycle Assessment 10(1):3–9.

Fuller, SK, Petersen, SR. 1996. Life cycle costing manual for the federal energy management program. NIST Handbook 135. Washington (DC): Government Printing Office.

Gabriel, R, Stichling, J, Jenka, B, Buxmann, K. 2003. Messung der Umweltleistung: integration von GaBi 4 in das Umweltdatenbank-System der Alcan. In: Spath, D, Lang, C, editors. Stoffstrommanagement: Entscheidungsunterstützung durch Umweltinformationen in der betrieblichen IT. Stuttgart (Germany): Fraunhofer Insitute IAO.

Galtung, J. 1996. Peace by peaceful means. London (UK): Sage.

Ganzheitlichen Bilanzierung. 1996. GaBi 2.0. Software zur Ganzheitlichen Bilanzierung, IKP der Universität Stuttgart, PE Product Engineering GmbH, Dettingen/Teck. Stuttgart (Germany): IKP der Universität Stuttgart.

Gensch, C-O, Grießhammer, R. 2004. PROSA — PKW-Flotte. Freiburg (Germany): Öko-Institut.

German EPA. 1991. Prognos/ISI, Konsistenzprüfung einer denkbaren zukünftigen Wasserstoffwirtschaft. Berlin (Germany): German EPA.

German Facility Management Association (GEFMA). 2006. GEFMA 220, Lebenszykluskostenrechnung im FM — Grundlagen und Anwendung, GEFMA e.V. Deutscher Verband für Facility Management. Bonn (Germany): German Facility Management Association.

Gesellschaft für umfassende Analysen GmbH (GUA). 2000. The benefits of using plastics. Vienna (Austria): GUA.

Gesellschaft für umfassende Analysen GmbH (GUA). 2001a. Review of recent literature on prevention cost by Institute of Public Finance and Infrastructure Policy at the Vienna University of Technology and GUA. Vienna (Austria): GUA.

Gesellschaft für umfassende Analysen GmbH (GUA). 2001b. Waste or recovered fuel, cost-benefit analysis. Vienna (Austria): GUA.

Goedkoop, M, Spriensmaa, R. 1999. The Eco-indicator 99: a damage oriented method for life cycle impact assessment. Amersfoort (The Netherlands): PRé Consultants.

Granta Design. 2004. Update of CES materials ecoselector, Cambridge engineering selector tool developed by Granta and Cambridge University, Professor Mike Ashby. Cambridge (UK). www.grantadesign.com.

Gray, R, Bebbington, J, editors. 2001. Accounting for the environment. 2nd ed. London (UK): Sage.

Guinée, JB (final ed.), Gorée, M, Heijungs, R, Huppes, G, Kleijn, R, de Koning, A, van Oers, L, Wegener Sleeswijk, A, Suh, S, Udo de Haes, HA, de Bruijn, H, van Duin, R, Huijbregts, MAJ. 2002. Handbook on life cycle assessment: operational guide to the ISO standards. Dordrecht (The Netherlands): Kluwer.

Günther, T, Kriegbaum, C. 1999. Fallstudie Heller und Pfennig. In Baum, HG, Coenenberg, AG, Günther, E, editors: Betriebliche Umweltökonomie in Fällen, Vol. 1. Munich (Germany): Oldenbourg. p 267–286.

Handfield, RB, Melnyk, SA, Clantone RJ, Curkovic, S. 2001. Integrating environmental concerns into the design process: the gap between theory and practice. IEEE Transactions on Engineering Management 48(2):189–208.

Hansen, DR, Mowen, MM. 1997. Management accounting. 4th ed. Cincinnati (OH): South-Western.

Hauschild, M, Wenzel, H. 1998. Scientific background. Vol. 2 of Environmental assessment of products. London (UK): Chapman & Hall.

Heemskerk, B, Pistorio, P, Scicluna, M. 2002. Sustainable development reporting. Striking the balance. Geneva (Switzerland): WBCSD.

Heemstra, FL. 1992. Software cost estimation. Information and Software Technology 34(10):627–639.

Hellweg, S, Hofstetter, P, Thomas, B, Hungerbühler, K. 2003. Discounting and the environment: should current impacts be weighted differently than impacts harming future generations? International Journal of Life Cycle Assessment, 8, 8–18.

Hilton, RW, Maher, MW, Selto, FH. 2000. Cost management: strategies for business decisions. New York (NY): McGraw-Hill.

Howarth, RB, Norgaard, RB. 1995. Intergenerational choices under global environmental change. In: Bromley, DW, editor. Handbook of environmental economics. Oxford (UK): Blackwell. p 111–138.

Hunkeler, D. 1999. Ecometrics for life cycle management: a conflict between sustainable development and family values. International Journal of Life Cycle Assessment 4:291.

Hunkeler, D. 2001. Return on environment: addressing the need for normalization and validation in EcoMetrics. In Life cycle management conference 2001 proceedings, Copenhagen, Denmark (August 27–29):45.

Hunkeler, D. 2006. Societal life cycle assessment: a methodology and case study. International Journal of Life Cycle Assessment 11(6):371–382. http://dx.doi.org/10.1065/lca2006.

Hunkeler, D, Biswas, G. 2000. Return on environment: an objective indicator to validate life cycle assessments? International Journal of Life Cycle Assessment 5(6):358–362.

Hunkeler, D, Rebitzer, G. 2001. Life cycle costing: paving the road to sustainable development? (editorial). Gate to EHS: Life Cycle Management, April 18, ecomed. http://www.scientificjournals.com/ehs.

Hunkeler, D, Rebitzer, G. 2003. Life cycle costing: paving the road to sustainable development? International Journal of Life Cycle Assessment 8(2):109–110.

Hunkeler, D, Rebitzer, G. 2005. The future of life cycle assessment. International Journal of Life Cycle Assessment 10(5):305–308.

Hunkeler, D, Saur, K, Jensen, A, Christensen, K, Rebitzer, G, Finkbeiner, M, Schmidt, WP, Stranddorf, H. 2004. Life cycle management. Pensacola (FL): SETAC Press.

Hunt, R, Franklin, W. 1974. Resource and Environmental Profile Analysis of Nine Beverage Container Alternatives. US Environmental Protection Agency, 1974 (EPA Report 530/SW-91c, NTIS No. PB 253486/5wp).

Hunt, R, Franklin, W. 1996. Personal reflections on the origin and the development of LCA in the USA. International Journal of Life Cycle Assessment 1(1):4–7.

Hunt, RG, Boguski, TK, Weitz, K, Sharma, A. 1998. Case studies examining streamlining techniques. International Journal of Life Cycle Assessment 3(1):36–42.

Huppes, G. 1993. Macro-environmental policy: principles and design. Amsterdam (The Netherlands): Elsevier. 430 p.

Huppes, G, van Rooijen, M, Kleijn, R, Heijungs, R, de Koning, A, van Oers, L. 2004. Life cycle costing and the environment. Report of a project commissioned by the Ministry of VROM-DGM for the RIVM Expertise Centre LCA. Amsterdam (The Netherlands): Ministry of VROM-DGM.

IKP/PE, PE Europe Life Cycle Engineering and IKP University of Stuttgart. 2005. Gabi 4 software and database. Stuttgart (Germany): IKP/PE. http://www.gabi-software.de (accessed June 2, 2005).

INFRAS/BEW. 1992. CO2 Perspektiven im Verkehr ($CO_2$ perspectives in traffic). Bern (Switzerland): Swiss Federal Institute for Energy.

International Electrotechnical Commission (IEC). 2004. IEC 60300-3-3, dependability management — part 3: application guide — section 3: life cycle costing. Geneva (Switzerland): International Electrotechnical Commission.

International Organization for Standardization (ISO). 2000–2001. International Standard ISO 15663: petroleum and natural gas industries — life cycle costing. Geneva (Switzerland): International Organization for Standardization.

International Standards Organization (ISO). 1997. International Standard ISO 14040: environmental management — life cycle assessment — principles and framework. Geneva (Switzerland): International Organization for Standardization.

International Standards Organization (ISO). 1998. International Standard ISO 14041: environmental management — life cycle assessment — goal and scope definition and inventory analysis. Geneva (Switzerland): International Organization for Standardization.

International Standards Organization (ISO). 2000a. International Standard ISO 14042: environmental management — life cycle assessment — life cycle impact assessment. Geneva (Switzerland): International Organization for Standardization.

International Standards Organization (ISO). 2000b. International Standard ISO 14043: environmental management — life cycle assessment — life cycle interpretation. Geneva (Switzerland): International Organization for Standardization.

International Standards Organization (ISO). 2002. International Standard ISO 14062: environmental management —integrating environmental aspects into product design and development. Geneva (Switzerland): International Organization for Standardization.

International Standards Organization (ISO). 2006a. ISO 14040, International Standard ISO 14040: environmental management — life cycle assessment: principles and framework, October. Geneva (Switzerland): International Organization for Standardization.

International Standards Organization (ISO). 2006b. ISO 14044, International Standard ISO 14044: environmental management — life cycle assessment: requirements and guidelines, October. Geneva (Switzerland): International Organization for Standardization.

ISO/DTS 2004. ISO/DTS 21929: Sustainability in building construction — sustainability indicators — part 1: framework for the development of indicators for buildings. Draft. Geneva (Switzerland): International Standards Organization.

Jensen, HR. 2002. Staging political consumption: a discourse analysis of the Brent Spar conflict as recast by the Danish mass media. Journal of Retailing and Consumer Services, 10, 71–80.

Kerzner, H. 2001. Project management: a systems approach to planning, scheduling, and controlling. New York (NY): Wiley.

Klöpffer, W. 2003. Life-cycle based methods for sustainable product development. International Journal of Life Cycle Assessment 8(3):157–159.

Klöpffer, W. 2006. The role of SETAC in the development of LCA. International Journal of Life Cycle Assessment 11(special issue 1):116–122.

Klöpffer, W, Renner, I. 2007. Lebenszyklusbasierte Nachhaltigkeitsbewertung von Produkten, In: Technikfolgenabschätzung — Theorie und Praxis Nr. 2, 16. Jg. Dezember 2007, Forschungszentrum Karlsruhe in der Helmholtz-Gemeinschaft, Institut für Technikfolgenabschätzung und Sustemanalyse, Karlsruhe (Germany).

Klüppel, H-J. 2005. The revision of ISO 14040-3. International Journal of Life Cycle Assessment 10(3):165.

Konar, S, Cohen, MA. 1997. Information as regulation: the effect of community right to know laws on toxic emissions. Journal of Environmental Economics and Management 32:109–124.

Kondo, Y, Nakamura, S. 2004. Evaluating alternative life-cycle strategies for electrical appliances by the waste input-output model. International Journal of Life Cycle Assessment, 9(4):236–246.

Krcmar, H. 1999. Integration des Umweltmanagements in die Softwarelandschaft des Unternehmens. In: Bullinger, H-J, Jürgens, G, Rey, U, editors. Betriebliche Umweltinformationssysteme in der Praxis. Stuttgart (Germany): Fraunhofer IRB.

Krishnan, V, Ulrich, KT. 2001. Product development decisions: a review of the literature. Management Science 47(1):1–21.

Kromrey, H. 2002. Empirische Sozialforschung. Opladen (Germany): Leske + Budrich.

Kunst, H. 2003. Ökologische Optimierung von Substitutionsentscheidungen. Dissertation an der Fakultät III—Prozesswissenschaften der Technischen Universität Berlin. Berlin (Germany): Technischen Universität Berlin.

Labuschagne, C, Brent, AC. 2006. Social indicators for sustainable project and technology life cycle management in the process industry. International Journal of Life Cycle Assessment 11(1):3–15.

Little, I, Mirrlees, J. 1969. Manual of project analysis in developing countries, Vol. 1. Paris: Organisation for Economic Cooperation and Development.

Margni, M, Swarr, T, Hunkeler, D. 2005. Moving from life cycle analysis to life cycle action: modular life cycle assessment applied to design incorporating the analytical hierarchy process for estimating EI values below the threshold. Barcelona (Spain): LCM-2005.

Marsmann, M. 2000. The ISO 14040 family. International Journal of Life Cycle Assessment 5(6):317–318.

Matsuno, Y, Tahara, K, Inaba, A. 1996. Life cycle inventories of washing machines. Journal of the Japan Institute of Energy 75-12, 1050–1055 (in Japanese).

Meadows, DL, Randers, J, Meadows, D. 1972. The limits to growth: a report for the Club of Rome's project on the predicament of mankind. New York (NY): Universe.

Metzger-Groom, S. 2003. A comparison of global wash habits and processes. Presentation at 50th SEPAWA Congress, October 8.

Miller, R, Blair, P. 1985. Input-output analysis. Englewood Cliffs (NJ): Prentice Hall.

Mishan, EJ. 1975. Cost-benefit analysis. London (UK): Allen and Unwin.

Möller, A. 2000. Grundlagen stoffstrombasierter Betrieblicher Umweltinformationssysteme. Bochum (Germany): Projekt Verlag.sher.

Mosovsky, JA, Dickinson, DA, Morabito, JM. 2000. Creating competitive advantage through resource productivity, eco- efficiency, and sustainability in the supply chain. In: Proceedings of 2000 IEEE International Symposium on Electronics and the Environment, San Francisco (CA): Institute of Electrical ande Electronics Engineers, Inc.

Munthe, C. 1997. Etiska aspekter på jordbruk (In Swedish). In: Jordbruksverkets rapport 1997:14. Jönköping (Sweden): Statens jordbruksverk.

Nakamura, S. 2003. The waste input-output table for Japan 1995, v. 2.2. http://www.f.waseda.jp/nakashin/research.html.

Nakamura, S, Kondo, Y. 2002. Input-output analysis of waste management. Journal of Industrial Ecology 6–1.

Nakamura, S, Kondo, Y. 2005. A waste input-output life-cycle cost analysis of the recycling of end-of-life electrical home appliances. Ecological Economics 57(3):494–506.

Norris, GA. 2001. Integrating life cycle cost analysis and LCA. International Journal of Life Cycle Assessment 6(2):118–120.

Norris, GA. 2006. Social impacts in product life cycles: towards life cycle attribute assessment. International Journal of Life Cycle Assessment 11(special issue 1):97–104.

Norris, GA, Laurin, L. 2004. Total cost accounting. Working paper, SETAC Life Cycle Costing Working Group. Pensacola (FL): SETAC.

Notarnicola, B, Tassielli, G, Nicoletti, GM. 2003. LCC and LCA of extra-virgin olive oil: organic vs. conventional. Paper presented at the international conference "Life Cycle Assessment in the Agri-Food Sector," Horsens, Denmark, October 6–8.

O'Brian, M, Doig, A, Clift, R. 1996. Social and environmental life cycle assessment (SELCA). International Journal of Life Cycle Assessment 1(4):231–237.

Office of Technology Assessment (OTA). 1996. The OTA legacy: 1972–1995. http://www.wws.princeton.edu/~ota/.

Okada, Y, Ueno, T, Onishi, H, Tachibana, H. 2002. Comparison of eco-efficiencies between several product categories of lamps. In: Proceedings of the Fifth International Conference on EcoBalance, November 6–8, Tsukuba, Japan.

Öko-Institut. 1987. Projektgruppe Ökologische Wirtschaft: Produktlinienanalyse (PLA). Freiburg (Germany): Kölner Volksblatt Verlag.

Öko-Institut. 2006. EcoTopTen-Kriterien für die Pkw-Flotte, Stand: 21. February. Freiburg (Germany): Öko-Institut.

Park, P-J, Tahara, K, Itsubo, N, Inaba, A. 2006. Estimation of product sustainability by combining quality, environmental and economic aspects. In: Proceedings of the Sixth International Conference on Ecobalance, October 25–27, Tsukuba, Japan. p 217–200.

Pearce, DW. 1983. Cost-benefit analysis. London (UK): Macmillan.

Pelzeter, A. 2005. Life cycle costs as a benchmark. Paper presented at the ERES Conference, Dublin, http://www.pelzeter.de/fileadmin/user_upload/apelzeter_upload/ERES_05_benchmark_paper_ape.pdf.

Pelzeter, A. 2006. Lebenszykluskosten von Immobilien — Einfluss von Lage, Gestaltung und Umwelt. Cologne (Germany): Rudolf Müller Verlag.

Potting, J, Hauschild, MZ. 2006. Spatial differentiation in life cycle impact assessment: a decade of method development to increase the environmental realism of LCIA. International Journal of Life Cycle Assessment 11(special issue 1):11–13.

PRé Consultants. 2004. SimaPro 6. LCA software. Amersfort (The Netherlands): PRé Consultants. http://www.pre.nl (accessed June 2, 2005).

Rebitzer, G. 2002. Integrating life cycle costing and life cycle assessment for managing costs and environmental impacts in supply chains. In: Seuring, S, Golbach, M, editors. Cost management in supply chains. Heidelberg (Germany): Physica-Verlag. p 127–146.

Rebitzer, G. 2005. Enhancing the application efficiency of life cycle assessment for industrial uses. PhD thesis, Swiss Federal Institute of Technology Lausanne. http://library.epfl.ch/theses/?nr=3307.

Rebitzer, G, Ekvall, T, editors. 2004. Scenario development in LCA. Pensacola (FL): Society for Environmental Toxicology and Chemistry (SETAC).

Rebitzer, G, Ekvall, T, Frischknecht, R, Hunkeler, D, Norris, G, Schmidt, W-P, Suh, S, Weidema, BP, Pennington, DW. 2004. Life cycle assessment (part 1): framework, goal & scope definition, inventory analysis, and applications. Environment International 30:701–720.

Rebitzer, G, Hunkeler, D. 2001. Merging economic and environmental information in life cycle management. In: Christiansen, S, Horup, M, Jensen, AA, editors. LCM 2001: proceedings. Copenhagen (Denmark): dk-TEKNIK. p 45–47.

Rebitzer, G, Hunkeler, D. 2003. Life cycle costing in LCM: ambitions, opportunities, and limitations, discussing a framework. International Journal of Life Cycle Assessment 8(5):253–256.

Rebitzer, G, Hunkeler, D, Braune, A, Stoffregen, A, Jolliet, O. 2002. Life cycle assessment of wastewater treatment options. In: Proceedings of the Fifth International Conference on Ecobalances, November 6–8, Tsukuba, Japan.

Rebitzer, G, Hunkeler, D, Jolliet, O. 2003. LCC — the economic pillar of sustainability: methodology and application to wastewater treatment. Environmental Progress 22(4):241–249.

Rebitzer, G, Hunkeler, D, Lichtenvort, K. 2004. Towards a code of practice for life cycle costing: results from the SETAC working group on LCC. In: Sixth International Conference of EcoBalance. Reprints, October 25–27, Tsukuba, Japan.

Riezler, S. 1996. Lebenszyklusrechnung: Instrument des Controlling strategischer Projekte. Wiesbaden (Germany): Gabler Verlag.

Ross, S, Evans, D, Webber, M. 2002. How LCA studies deal with uncertainty. LCA 7(1):47–52.

Royston, M. 1979. Pollution prevention pays. Oxford (UK): Pergamon.

Roztocki, N. 1998. Introduction to activity based costing (ABC). Pittsburgh (PA): University of Pittsburgh. http://www.pitt.edu/~roztocki/abc/abctutor/.

Rüdenauer, I, Gensch, C-O. 2005a. Eco-efficiency analysis of washing machines. Refinement of Task 4: further use versus substitution of washing machines in stock. Freiburg (Germany): Öko-Institut.

Rüdenauer, I, Gensch, C-O. 2005b. Environmental and economic evaluation of accelerated replacement of domestic appliances: case study refrigerators and freezers. Freiburg (Germany): Öko-Institut.

Rüdenauer, I, Gensch, C-O, Quack, D. 2004. Eco-efficiency analysis of washing machines: life cycle assessment and determination of optimal life span. Freiburg (Germany): Öko-Institut.

Rüdenauer, I, Grießhammer, R. 2004. PROSA — Waschmaschinen. Produkt-Nachhaltigkeitsanalyse von Waschmaschinen und Waschprozessen (PROSA — washing machines. Product sustainability assessment of washing machines and washing processes). Freiburg (Germany): Öko-Institut.

Saling, P, Kicherer, A, Dittrich-Krämer, B, Wittlinger, R, Zombik, W, Schmidt, I, Schrott, W, Schmidt, S. 2002. Eco-efficiency analysis by BASF: the method. International Journal of Life Cycle Assessment 7(4):203–218.

Schmidt, W-P. 2003. Life cycle costing as part of design for environment: environmental business cases. International Journal of Life Cycle Assessment 8(3):167–174.

Seattle. 2003. City of Seattle: water meter reading with segues: life cycle cost analysis report. Seattle (WA): City of Seattle.

Sen, A. 1987. On Ethics and Economics. Oxford and New York: Basil Blackwell.

Seuring, S. 2002. Supply chain costing: a conceptual framework, In: Seuring, S, Goldbach, M, editors. Cost management in supply chains. Heidelberg (Germany): Physica. p 15–30.

Seuring, S. 2003. Cost management in the textile chain: reducing environmental impacts and costs for green products. In: Bennett, S, Schaltegger, S, Rikhardsson, P, editors. Environmental management accounting: purpose and progress. Dordrecht (The Netherlands): Kluwer. p 233–256.

Shapiro, KG. 2001. Incorporating costs in LCA. International Journal of Life Cycle Assessment 6(2):121–123.

Sherif, YS, Kolarik, WJ. 1981. Life cycle costing: concept and practice. OMEGA: The International Journal of Management Science 9(3):287–296.

Singer, P. 1975. Animal liberation. New York (NY): HarperCollins.

Slagmulder, R. 2002. Managing costs across the supply chain. In: Seuring, S, Goldbach, M. Cost management in supply chains. Heidelberg (Germany): Physica Verlag. p 75–88.

Smulders, E. 2002. Laundry detergents. Weinheim (Germany): Wiley-VCH.

Society of Automotive Engineers (SAE). 1992. Aerospace recommended practice (ARP4293), life cycle cost — techniques and applications. Warrendale (PA): Society of Automotive Engineers.

Society of Environmental Toxicology and Chemistry (SETAC). 1993. Guidelines for life-cycle assessment: a "code of practice." Based on a Workshop at Sesimbra, Portugal, March 31–April 3, 1993. Pensacola (FL): SETAC.

Sonnemann, G, Castells, F, Schumacher, M. 2003. Integrated life-cycle and risk assessment for industrial processes. Boca Raton (FL): CRC Press.

Spash, CL. 1997. Ethics and environmental attitudes with implications for economic valuation. Journal of Environmental Management 50:403–416.

Standard & Poor's. 2004. Standard & Poor's structured finance ratings, real estate finance, environmental criteria. New York (NY): Standard & Poor's.

Standards Australia and Standards New Zealand (AS/NZS). 1999. AS/NZS 4536: Standards Australia and Standards New Zealand: life cycle costing — an application guide. Canberra (Australia): Standards Australia and Standards New Zealand.

Steen, B. 1999a. A systematic approach to environmental priority strategies in product development (EPS): version 2000 — general system characteristics. Göteborg (Sweden): Chalmers University of Technology, Center for Environmental Assessment of Products and Material Systems (CPM). 65 p.

Steen, B. 1999b. A systematic approach to environmental priority strategies in product development (EPS): version 2000 — models and data. Göteborg (Sweden): Chalmers University of Technology, Center for Environmental Assessment of Products and Material Systems (CPM). 312 p.

Steen B, et al. 2004. Development of interpretation keys for environmental product declarations (EPD). Submitted to Journal of Cleaner Production, May 2004.

Stier, W. 1999. Empirische Forschungsmethoden. Heidelberg (Germany): Springer.

Stirling, A. 1997. Limits to the value of external costs. Energy / Policy 25:517–540.

Stoeckl, N. 2004. The private costs and benefits of environmental self-regulation: which firms have most to gain? Business Strategy and the Environment 13:135–155.

Stolp, A. 2003. Citizen values assessment. In: Becker, HA, Vancla, F, editors. The international handbook of social impact assessment. Cheltenham (UK): Edward Elgar. p 231–257.

Suh, S, Huppes, G. 2005. Methods in life cycle inventory (LCI) of a product. Journal of Cleaner Production 13(7):687–697.

Suh, S, Lenzen, M, Treloar, G, Hondo, H, Horvath, A, Huppes, G, Jolliet, O, Klann, U, Krewitt, W, Moriguchi, Y, Munksgaard, J, Norris, G. 2004. System boundary selection in life-cycle inventories using hybrid approaches. Environmental Science & Technology 38(3):657–664.

Tinbergen, J. 1961. De econonische balans van het Deltaplan [The economic balance of the Deltaplan]. Report Deltacommisie. Den Haag (the Netherlands): Staatsdrukkerij.

Töpfer, K. 2002. The launch of the UNEP-SETAC Life Cycle Initiative (Prague, April 28, 2002) (editorial). International Journal of Life Cycle Assessment 7(4):191.

Udo de Haes, HA, et al. 2002. Life cycle impact assessment: striving towards best practice. Pensacola (FL): SETAC.

Udo de Haes, HA, Heijungs, R, Suh, S, Huppes, G. 2004. Three strategies to overcome the limitations of life-cycle assessment. Journal of Industrial Ecology 8(3):19–32.

Ullman, DG. 2001. Robust decision-making for engineering design. Journal of Engineering Design 12(1):3–13.

UNEP. 1992. United Nations Environment Programme: Agenda 21, revealed and adopted by 179 governments at the 1992 United Nations Conference on Environment and Development (Earth Summit), Rio de Janeiro (Brazil).

UNEP–SETAC. 2005. Life cycle initiative. http://www.uneptie.org/pc/sustain/lcinitiative/home.htm (accessed August 25, 2005).

United Nations. 2005. UN millennium development goals. http://www.un.org/millenniumgoals/ (accessed August 25, 2005).

United Nations Statistics Division. 2007. UN classifications registry. http://unstats.un.org/unsd/cr/registry/.

US Department of Defense (DoD). 1973. Life cycle costing guide for system acquisitions, January. Washington (DC): Author.

US Department of Defense (DoD). 1999. Parametric estimating handbook. 2nd ed. Washington (DC): Author, app. A.

US Department of Energy (DoE). 2005. Gasoline prices, national and regional prices. http://tonto.eia.doe.gov/oog/info/gdu/gasdiesel.asp.

US Department of Transportation, Federal Highway Administration. Life-cycle cost analysis in pavement design. Pavement Division Interim Technical Bulletin, September 1998.

US Environmental Protection Agency (USEPA). 1989. Toxic release inventory, June 19. Washington (DC): US Environmental Protection Agency.

US Environmental Protection Agency (USEPA). 1996. Valuing potential environmental liabilities for managerial decision-making: a review of available techniques. EPA report EPA 742-R-96-003. Washington (DC): USEPA, Office of Pollution Prevention and Toxics.

US Environmental Protection Agency (USEPA). 1999. OAQPS economic analysis resource document. Washington (DC): US Environmental Protection Agency, Office of Air Quality Planning & Standards, Innovative Strategies and Economics Group. http://www.epa.gov/ttn/ecas/econdata/Rmanual2/index.html.

Vanclay, F. 2003. Conceptual and methodological advances in social impact assessment. In: Becker, HA, Vanclay, F, editors. The international handbook of social impact assessment. Cheltenham (UK): Edward Elgar. p 1–9.

van Schooten, M, Vanclay, F, Stootweg, R. 2003. Conceptualizing social change processes and social impacts, In: Becker, HA, Vanclay, F, editors. The international handbook of social impact assessment. Cheltenham (UK): Edward Elgar. p 74–91.

Verein Deutscher Ingenieure (VDI; Society of German Engineers). 1984. VDI-Richtlinie 2225, Technisch-wirtschaftliches Konstruieren, VDI — Gesellschaft Entwicklung Konstruktion und Vertrieb. Düsseldorf (Germany): VDI.

von Weizsäcker, E, Lovins, AB, Lovins, LH. 1998. Factor 4: doubling wealth, halving resource use — a report to the Club of Rome. London (UK): Earthscan.

Weidema, BP. 2006. The integration of economic and social aspects in life cycle impact assessment. International Journal of Life Cycle Assessment 11(special issue 1):89–96.

Wenzel, H, Hauschild, M, Alting, L. 1997. Methodology, tools and case studies in product development. Vol. 1 of Environmental assessment of products. London (UK): Chapman & Hall.

White, AL, Savage, D, Shapiro, KG. 1996. Life-cycle costing: concepts and application. In: Curran, MA, editor. Environmental life-cycle assessment. New York (NY): McGraw-Hill, 7.1 to 7.19.

Wicke, L. 1992. Betriebliche Umweltökonomie — Eine praxisorientierte Einführung. Munich (Germany): Franz Vahlen.

Willard, B. 2002. The sustainability advantage: seven business case benefits of a triple bottom line. Gabriola Islands (BC, Canada): New Society.

World Business Council for Sustainable Development (WBCSD). 2003. Eco-efficiency: creating more value with less impact. Geneva (Switzerland): World Business Council for Sustainable Development.

Zehbold, C. 1996. Lebenszykluskostenrechnung. Wiesbaden (Germany): Gabler Verlag.

# Index

## A



## B



## C

## D



## E

## N



## O



## P



## R



## S



## T

## U



## V



## W

## Other Titles from the Society of Environmental Toxicology and Chemistry (SETAC)

*Assessing the Hazard of Metals and Inorganic Metal Substances in Aquatic and Terrestrial Systems*
Adams and Chapman, editors
2006

*Perchlorate Ecotoxicology*
Kendall and Smith, editors
2006

*Natural Attenuation of Trace Element Availability in Soils*
Hamon, McLaughlin, Stevens, editors
2006

*Mercury Cycling in a Wetland-Dominated Ecosystem: A Multidisciplinary Study*
O'Driscoll, Rencz, Lean
2005

*Atrazine in North American Surface Waters: A Probabilistic Aquatic Ecological Risk Assessment*
Giddings, editor
2005

*Effects of Pesticides in the Field*
Liess, Brown, Dohmen, Duquesne, Hart, Heimbach, Kreuger, Lagadic, Maund, Reinert, Streloke, Tarazona
2005

*Human Pharmaceuticals: Assessing the Impacts on Aquatic Ecosystems*
Williams, editor
2005

*Toxicity of Dietborne Metals to Aquatic Organisms*
Meyer, Adams, Brix, Luoma, Stubblefield, Wood, editors
2005

*Toxicity Reduction and Toxicity Identification Evaluations for Effluents, Ambient Waters, and Other Aqueous Media*
Norberg-King, Ausley, Burton, Goodfellow, Miller, Waller, editors
2005

*Use of Sediment Quality Guidelines and Related Tools for the Assessment of Contaminated Sediments*
Wenning, Batley, Ingersoll, Moore, editors
2005

*Life-Cycle Assessment of Metals*
Dubreuil, editor
2005

*Working Environment in Life-Cycle Assessment*
Poulsen and Jensen, editors
2005

*Life-Cycle Management*
Hunkeler, Saur, Rebitzer, Finkbeiner, Schmidt, Jensen, Stranddorf, Christiansen
2004

*Scenarios in Life-Cycle Assessment*
Rebitzer and Ekvall, editors

*Ecological Assessment of Aquatic Resources: Linking Science to Decision-Making*
Barbour, Norton, Preston, Thornton, editors
2004

*Life-Cycle Assessment and SETAC: 1991–1999*
15 LCA publications on CD-ROM
2003

*Amphibian Decline: An Integrated Analysis of Multiple Stressor Effects*
Greg Linder, Sherry K. Krest, Donald W. Sparling
2003

*Metals in Aquatic Systems:*
*A Review of Exposure, Bioaccumulation, and Toxicity Models*
Paquin, Farley, Santore, Kavvadas, Mooney, Winfield, Wu, Di Toro
2003

*Silver: Environmental Transport, Fate, Effects, and Models:*
*Papers from Environmental Toxicology and Chemistry, 1983 to 2002*
Gorusch, Kramer, La Point
2003

*Code of Life-Cycle Inventory Practice*
de Beaufort-Langeveld, Bretz, van Hoof, Hischier, Jean, Tanner, Huijbregts, editors
2003

*Contaminated Soils: From Soil–Chemical Interactions to Ecosystem Management*
Lanno, editor
2003

*Environmental Impacts of Pulp and Paper Waste Streams*
Stuthridge, van den Heuvel, Marvin, Slade, Gifford, editors
2003

*Life-Cycle Assessment in Building and Construction*
Kotaji, Edwards, Shuurmans, editors
2003

*Porewater Toxicity Testing: Biological, Chemical, and Ecological Considerations*
Carr and Nipper, editors
2003

*Reevaluation of the State of the Science for Water-Quality Criteria Development*
Reiley, Stubblefield, Adams, Di Toro, Erickson, Hodson, Keating Jr, editors
2003

*Community-Level Aquatic System Studies—Interpretation Criteria (CLASSIC)*
Giddings, Brock, Heger, Heimbach, Maund, Norman, Ratte,
Schäfers, Streloke, editors
2002

*Interconnections between Human Health and Ecological Variability*
Di Giulio and Benson, editors
2002

*Life-Cycle Impact Assessment: Striving towards Best Practice*
Udo de Haes, Finnveden, Goedkoop, Hauschild, Hertwich, Hofstetter, Jolliet,
Klöpffer, Krewitt, Lindeijer, Müller-Wenk, Olsen, Pennington, Potting, Steen, editors
2002

*Silver in the Environment: Transport, Fate, and Effects*
Andren and Bober, editors
2002

*Test Methods to Determine Hazards for Sparingly Soluble Metal Compounds in Soils*
Fairbrother, Glazebrook, van Straalen, Tararzona, editors
2002

*9th LCA Case Studies Symposium*
2001

*Avian Effects Assessment: A Framework for Contaminants Studies*
Hart, Balluff, Barfknecht, Chapman, Hawkes, Joermann, Leopold, Luttik, editors
2001

*Bioavailability of Metals in Terrestrial Ecosystems:*
*Importance of Partitioning for Bioavailability to Invertebrates, Microbes, and Plants*
Allen, editor
2001

*Ecological Variability: Separating Natural from Anthropogenic Causes of Ecosystem Impairment*
Baird and Burton, editors
2001

*Guidance Document on Regulatory Testing and Risk Assessment Procedures for Protection Products with Non-Target Arthropods (ESCORT 2)*
Candolfi, Barrett, Campbell, Forster, Grady, Huet, Lewis, Schmuck, Vogt, editors
2001

*Impacts of Low-Dose, High-Potency Herbicides on Nontarget and Unintended Plant Species*
Ferenc, editor
2001

*Risk Management: Ecological Risk-Based Decision-Making*
Stahl, Bachman, Barton, Clark, deFur, Ells, Pittinger, Slimak, Wentsel, editors
2001

*8th LCA Case Studies Symposium*
2000

*Development of Methods for Effects-Driven Cumulative Effects Assessment Using Fish Populations: Moose River Project*
Munkittrick, McMaster, Van Der Kraak, Portt, Gibbons, Farwell, Gray, authors
2000

*Ecotoxicology of Amphibians and Reptiles*
Sparling, Linder, Bishop, editors
2000

*Environmental Contaminants and Terrestrial Vertebrates:*
*Effects on Populations, Communities, and Ecosystems*
Albers, Heinz, Ohlendorf, editors
2000

*Evaluation of Persistence and Long-Range Transport of Organic Chemicals in the Environment*
Klečka, Boethling, Franklin, Grady, Graham, Howard, Kannan, Larson, Mackay, Muir, van de Meent, editors
2000

*Multiple Stressors in Ecological Risk and Impact Assessment: Approaches to Risk Estimation*
Ferenc and Foran, editors
2000

*Natural Remediation of Environmental Contaminants:*
*Its Role in Ecological Risk Assessment and Risk Management*
Swindoll, Stahl, Ells, editors
2000

*7th LCA Case Studies Symposium*
1999

*Evaluating and Communicating Subsistence Seafood Safety in a Cross-Cultural Context:*
*Lessons Learned from the* Exxon Valdez *Oil Spill*
Field, Fall, Nighswander, Peacock, Varanasi, editors
1999

*Ecotoxicology and Risk Assessment for Wetlands*
Lewis, Mayer, Powell, Nelson, Klaine, Henry, Dickson, editors
1999

*Endocrine Disruption in Invertebrates: Endocrinology, Testing, and Assessment*
DeFur, Crane, Ingersoll, Tattersfield, editors
1999

*Guidance Document on Higher-Tier Aquatic Risk Assessment for Pesticides (HARAP)*
Campbell, Arnold, Brock, Grandy, Heger, Heimbach, Maund, Streloke, editors
1999

*Linkage of Effects to Tissue Residues: Development of a Comprehensive Database for Aquatic Organisms Exposed to Inorganic and Organic Chemicals*
Jarvinen and Ankley, editors
1999

*Multiple Stressors in Ecological Risk and Impact Assessment*
Foran and Ferenc, editors
1999

*Reproductive and Developmental Effects of Contaminants in Oviparous Vertebrates*
Di Giulio and Tillitt, editors
1999

*Restoration of Lost Human Uses of the Environment*
Grayson Cecil, editor
1999

*6th LCA Case Studies Symposium*
1998

*Advances in Earthworm Ecotoxicology*
Sheppard, Bembridge, Holmstrup, Posthuma, editors
1998

*Ecological Risk Assessment: A Meeting of Policy and Science*
Peyster and Day, editors
1998

*Ecological Risk Assessment Decision-Support System: A Conceptual Design*
Reinert, Bartell, Biddinger, editors
1998

*Ecotoxicological Risk Assessment of the Chlorinated Organic Chemicals*
Carey, Cook, Giesy, Hodson, Muir, Owens, Solomon, editors
1998

*Principles and Processes for Evaluating Endocrine Disruption in Wildlife*
Kendall, Dickerson, Giesy, Suk, editors
1998

*Radiotelemetry Applications for Wildlife Toxicology Field Studies*
Brewer and Fagerstone, editors
1998

*Sustainable Environmental Management*
Barnthouse, Biddinger, Cooper, Fava, Gillett, Holland, Yosie, editors
1998

*Uncertainty Analysis in Ecological Risk Assessment*
Warren-Hicks and Moore, editors
1998

*5th LCA Case Studies Symposium*
1997

*Atmospheric Deposition of Contaminants to the Great Lakes and Coastal Waters*
Baker, editor
1997

*Biodegradation Kinetics: Generation and Use of Data for Regulatory Decision-Making*
Hales, Feijtel, King, Fox, Verstraete, editors
1997

*Biotransformation in Environmental Risk Assessment*
Sijm, de Bruijn, de Boogt, de Wolf, editors
1997

*Chemical Ranking and Scoring: Guidelines for Relative Assessments of Chemicals*
Swanson and Socha, editors
1997

*Chemically Induced Alterations in Functional Development and Reproduction of Fishes*
Rolland, Gilbertson, Peterson, editors
1997

*Ecological Risk Assessment of Contaminated Sediments*
Ingersoll, Dillon, Biddinger, editors
1997

*Life-Cycle Impact Assessment: The State-of-the-Art,* 2nd ed.
Barnthouse, Fava, Humphreys, Hunt, Laibson, Moesoen, Owens, Todd, Vigon, Weitz, Young, editors
1997

*Public Policy Application of Life-Cycle Assessment*
Allen and Consoli, editors
1997

*Quantitative Structure-Activity Relationships (QSAR) in Environmental Sciences* VII
Chen and Schüürmann, editors
1997

*Reassessment of Metals Criteria for Aquatic Life Protection: Priorities for Research and Implementation*
Bergman and Dorward-King, editors
1997

*Simplifying LCA: Just a Cut?*
Christiansen, editor
1997

*Workshop of Endocrine Modulators and Wildlife: Assessment and Testing (EMWAT)*
Tattersfield, Matthiessen, Campbell, Grandy, Länge, editors
1997

*Asking the Right Questions: Ecotoxicology and Statistics*
Chapman, Crane, Wiles, Noppert, McIndoe, editors
1996

*Pesticides, Soil Microbiology and Soil Quality*
Anderson, Arnold, Malkomes, Lagacherie, Oliveira, Plicken, Tarry, Soulas, Torstensson, editors
1996

*Towards a Methodology for Life-Cycle Impact Assessment*
Udo de Haes, editor
1996

*Whole Effluent Toxicity Testing: An Evaluation of Methods and Prediction of Receiving System Impacts*
Grothe, Dickson, Reed-Judkins, editors
1996

*Guidance Document on Regulatory Testing Procedures for Pesticides with Non-Target Arthropods*
Barrett, Grady, Harrison, Hassan, Oomen, editors
1995

*Procedures for Assessing the Environmental Fate and Ecotoxicity of Pesticides*
Lynch, editor
1995

*The Multi-Media Fate Model: A Vital Tool for Predicting the Fate of Chemicals*
Cowan, D. Mackay, Feijtel, Meent, Di Guardo, Davies, N. Mackay, editors
1995

*Allocation in LCA*
Huppes, Schneider, editors
1994

*Aquatic Dialogue Group: Pesticide Risk Assessment and Mitigation*
Baker, Barefoot, Beasley, Burns, Caulkins, Clark, Feulner, Giesy, Graney, Griggs, Jacoby, Laskowski, Maciorowski, Mihaich, Nelson, Parrish, Siefert, Solomon, van der Schalie, editors
1994

*Integrating Impact Assessment into LCA*
Udo de Haes, Jensen, Klöpffer, Lindfors, editors
1994

*Life-Cycle Assessment Data Quality: A Conceptual Framework*
Fava, Jensen, Lindfors, Pomper, De Smet, Warren, Vigon, editors
1994

*Sediment Toxicity Tests and Bioassays for Freshwater and Marine Environments*
Hill, Matthiessen, Heimbach, editors
1994

*A Conceptual Framework for Life-Cycle Impact Assessment*
Fava, Consoli, Denison, Dickson, Mohin, Vigon, editors
1993

*Environmental Modelling: The Next Ten Years*
Stebbing, Travis, Matthiessen, editors
1993

*Guidelines for Life-Cycle Assessment: A "Code of Practice"*
Consoli, Allen, Boustead, Fava, Franklin, Jensen, Oude, Parrish, Perriman, Postlethewaite, Quay, Seguin, Vigon, editors
1993

*Life-Cycle Assessment: Inventory, Classification, Valuation, Data Bases*
1992

*Workshop on Aquatic Microcosms for Ecological Assessment of Pesticides*
1992

*Guidance Document on Testing Procedures for Pesticides in Mesocosms*
1991

*A Technical Framework for Life-Cycle Assessment*
Fava, Denison, Jones, Curran, Vigon, Selke, Barnum, editors
1991

*Research Priorities in Environmental Risk Assessment*
Fava, Adams, Larson, Dickson, Dickson, Bishop, editors
1987

# SETAC

A Professional Society for Environmental Scientists and Engineers and Related Disciplines Concerned with Environmental Quality

The Society of Environmental Toxicology and Chemistry (SETAC), with offices currently in North America and Europe, is a nonprofit, professional society established to provide a forum for individuals and institutions engaged in the study of environmental problems, management and regulation of natural resources, education, research and development, and manufacturing and distribution.

Specific goals of the society are

- Promote research, education, and training in the environmental sciences.
- Promote the systematic application of all relevant scientific disciplines to the evaluation of chemical hazards.
- Participate in the scientific interpretation of issues concerned with hazard assessment and risk analysis.
- Support the development of ecologically acceptable practices and principles.
- Provide a forum (meetings and publications) for communication among professionals in government, business, academia, and other segments of society involved in the use, protection, and management of our environment.

These goals are pursued through the conduct of numerous activities, which include:

- Hold annual meetings with study and workshop sessions, platform and poster papers, and achievement and merit awards.
- Sponsor a monthly scientific journal, a newsletter, and special technical publications.
- Provide funds for education and training through the SETAC Scholarship/Fellowship Program.
- Organize and sponsor chapters to provide a forum for the presentation of scientific data and for the interchange and study of information about local concerns.
- Provide advice and counsel to technical and nontechnical persons through a number of standing and ad hoc committees.

SETAC membership currently is composed of more than 5000 individuals from government, academia, business, and public-interest groups with technical backgrounds in chemistry, toxicology, biology, ecology, atmospheric sciences, health sciences, earth sciences, and engineering.

If you have training in these or related disciplines and are engaged in the study, use, or management of environmental resources, SETAC can fulfill your professional affiliation needs.

All members receive a newsletter highlighting environmental topics and SETAC activities, and reduced fees for the Annual Meeting and SETAC special publications.

All members except Students and Senior Active Members receive monthly issues of Environmental Toxicology and Chemistry (ET&C) and Integrated Environmental Assessment and Management (IEAM), peer-reviewed journals of the Society. Student and Senior Active Members may subscribe to the journal. Members may hold office and, with the Emeritus Members, constitute the voting membership.

If you desire further information, contact the appropriate SETAC Office.

1010 North 12th Avenue
Pensacola, Florida 32501-3367 USA
T 850 469 1500 F 850 469 9778
E setac@setac.org

Avenue de la Toison d'Or 67
B-1060 Brussels, Belgium
T 32 2 772 72 81 F 32 2 770 53 86
E setac@setaceu.org

www.setac.org
Environmental Quality Through Science®